The Cambridge Handbook of Earth Scie

This handbook presents an indispensabl
figures about the Earth. It brings together reliable physical, and
and historical data in a series of 145 easy-to-read tables, supplemented by
maps, charts and colour plates. Eleven chapters cover topics spanning the
Earth's geosphere, hydrosphere, atmosphere and biosphere, with one chapter
focusing on other bodies in the Solar System. Full references for the original
data sources are provided to enable users to access further detail and, where
relevant, contentious or alternative data are indicated in supplementary notes.
The appendix provides practical information on units and conversion factors.

Compact and easy to use, this handy book provides a time-saving first
point of reference for researchers, students and practitioners in the Earth and
Environmental Sciences. It allows scientists easy access to basic information on
topics outside their specialisation, and is also a convenient resource for
non-scientists such as economists, policy makers and journalists.

PAUL HENDERSON is an Honorary Professor in the Department of Earth
Sciences at University College London and was previously Head of the
Department of Mineralogy and Director of Science at the Natural History
Museum. He was President of the Mineralogical Society of Great Britain and
Northern Ireland (1990–1992) and Vice President of the Geological Society of
London (2002–2008). Professor Henderson's research interests focus on the
geochemistry of igneous and related rocks, and he is the author of the textbook
Inorganic Geochemistry (1982) and the editor of *Rare Earth Element
Geochemistry* (1984).

GIDEON HENDERSON is Professor of Earth Sciences at Oxford University and
was previously an Associate Research Scientist at the Lamont Doherty Earth
Observatory of Columbia University, New York. His research interests involve
the application of geochemical tools to understand the Earth's surface
environment including climate change, past ocean circulation and geochemical
fluxes from land to sea. Professor Henderson is also an editor of
Uranium-Series Geochemistry (2003).

The Cambridge Handbook of Earth Science Data

Paul Henderson

Earth Sciences, University College London,
and Department of Mineralogy, The Natural History Museum, London

Gideon M. Henderson

Department of Earth Sciences, University of Oxford
and University College, Oxford

CAMBRIDGE
UNIVERSITY PRESS

CAMBRIDGE UNIVERSITY PRESS
Cambridge, New York, Melbourne, Madrid, Cape Town, Singapore,
São Paulo, Delhi

Cambridge University Press
The Edinburgh Building, Cambridge CB2 8RU, UK

Published in the United States of America by Cambridge University Press,
New York

www.cambridge.org
Information on this title: www.cambridge.org/9780521693172

First published 2009

Printed in the United Kingdom at the University Press, Cambridge

A catalogue record for this publication is available from the British Library

Library of Congress Cataloguing in Publication data
Henderson, Paul, 1940–
The Cambridge handbook of earth science data / Paul Henderson, Gideon
Henderson.
    p.   cm.
Includes index.
ISBN 978-0-521-69317-2
1. Earth sciences – Handbooks, manuals, etc.   I. Henderson, Gideon,
1968–   II. Title.
QE26.3H45   2009
550.21 – dc22   2009008252

ISBN 978-0-521-69317-2 paperback

Contents

Preface

This book aims to be an accessible and convenient source for basic data in the Earth Sciences. It is not only for use by students but also for instructors, researchers and other professional geoscientists who want rapid access to core information across some of the many areas – biological to cosmological – that comprise the Earth Sciences today. It is not intended as an in-depth research tool within any one topic but more of a starting-off point and it should be seen as complementary to other sources such as course textbooks.

The scope and design of the book are as close to a pocket book as they reasonably can be, so as to enable use of the book in many circumstances – the teaching laboratory, the conference or lecture hall, the study room, while travelling and even perhaps in the field. SI units are used for the most part but on occasion other units that still tend to be in common use are quoted (conversions can be found in the Appendix). The selection of material has been made in order to fulfil the aims while at the same time keeping the size within reasonable bounds. There is an emphasis on 'natural' processes rather than anthropogenic influences; error analysis is not generally covered, and numerical data are given precedence over diagrammatic representation. Information on data sources is provided to help the user pursue a topic further. Inevitably such a selection will be something of an experiment, so we look forward to suggestions from users on how it might be changed and improved should, at a later date, another edition be considered. We also welcome our attention being drawn to any errors that may have crept in despite our attempts to avoid them.

I have compiled most of the contents, with Gideon focusing on Chapters 4 (Aqueous Earth) and 5 (Gaseous Earth) together with his overview of the isotopic tables of Chapter 8 as well as contributing diagrams with isotopic data in Chapter 7 (Earth history). We very much hope that users of this book will find their need for a one-stop and straightforward data source satisfied – a need we felt and which prompted us to produce this book.

Paul Henderson
University College London

Acknowledgements

We thank the following people who kindly made suggestions, comments or reviews of parts of this book: Tim Atkinson, Paul Bown, John Brodholt, Michel Guiraud, Adrian Jones, Paul Kenrick, Conall MacNiocaill, John McArthur, David Manning, Bill McGuire, David Price, Phil Rainbow, Peter Rawson, Stuart Robinson, Nick Rogers, Sara Russell, Peter Sammonds, Andy Saunders, Chris Stringer, Claudio Vita-Finzi, Sarah Watkinson and Ian Wood. We also thank Susan Francis of Cambridge University Press for her help in guiding the production of this book.

1 The Solar System

Table 1.1 Solar System – elemental abundances

	Element	Log abundance [H] = 12	Atoms per 10^6 Si atoms
1	H	12	2.431×10^{10}
2	He	10.984	2.343×10^{9}
3	Li	3.35	55.47
4	Be	1.48	0.7374
5	B	2.85	17.32
6	C	8.46	7.079×10^{6}
7	N	7.90	1.950×10^{6}
8	O	8.76	1.413×10^{7}
9	F	4.53	841.1
10	Ne	7.95	2.148×10^{6}
11	Na	6.37	5.751×10^{4}
12	Mg	7.62	1.020×10^{6}
13	Al	6.54	8.410×10^{4}
14	Si	7.61	1×10^{6}
15	P	5.54	8.373×10^{3}
16	S	7.26	4.449×10^{5}
17	Cl	5.33	5.237×10^{3}
18	Ar	6.62	1.025×10^{5}
19	K	5.18	3.692×10^{3}
20	Ca	6.41	6.287×10^{4}
21	Sc	3.15	34.20
22	Ti	5.00	2.422×10^{3}
23	V	4.07	288.4
24	Cr	5.72	1.286×10^{4}
25	Mn	5.58	9.168×10^{3}
26	Fe	7.54	8.380×10^{5}
27	Co	4.98	2.323×10^{3}
28	Ni	6.29	4.780×10^{4}
29	Cu	4.34	527.0
30	Zn	4.70	1.226×10^{3}
31	Ga	3.17	35.97
32	Ge	3.70	120.6

Table 1.1 (*cont.*)

	Element	Log abundance [H] = 12	Atoms per 10^6 Si atoms
33	As	2.40	6.089
34	Se	3.43	65.79
35	Br	2.67	11.32
36	Kr	3.36	55.15
37	Rb	2.43	6.572
38	Sr	2.99	23.64
39	Y	2.28	4.608
40	Zr	2.67	11.33
41	Nb	1.49	0.755 4
42	Mo	2.03	2.601
44	Ru	1.89	1.900
45	Rh	1.18	0.370 8
46	Pd	1.77	1.435
47	Ag	1.30	0.491 3
48	Cd	1.81	1.584
49	In	0.87	0.181 0
50	Sn	2.19	3.733
51	Sb	1.14	0.329 2
52	Te	2.30	4.815
53	I	1.61	0.997 5
54	Xe	2.35	5.391
55	Cs	1.18	0.367 1
56	Ba	2.25	4.351
57	La	1.25	0.440 5
58	Ce	1.68	1.169
59	Pr	0.85	0.173 7
60	Nd	1.54	0.835 5
62	Sm	1.02	0.254 2
63	Eu	0.60	0.095 13
64	Gd	1.13	0.332 1

(*cont.*)

Table 1.1 (*cont.*)

	Element	Log abundance [H] = 12	Atoms per 10^6 Si atoms
65	Tb	0.38	0.059 07
66	Dy	1.21	0.386 2
67	Ho	0.56	0.089 86
68	Er	1.02	0.255 4
69	Tm	0.18	0.037 00
70	Yb	1.01	0.248 4
71	Lu	0.16	0.035 72
72	Hf	0.84	0.169 9
73	Ta	−0.06	0.020 99
74	W	0.72	0.127 7
75	Re	0.33	0.052 54
76	Os	1.44	0.673 8
77	Ir	1.42	0.644 8
78	Pt	1.75	1.357
79	Au	0.91	0.195 5
80	Hg	1.23	0.412 8
81	Tl	0.88	0.184 5
82	Pb	2.13	3.258
83	Bi	0.76	0.138 8
90	Th	0.16	0.035 12
91	U	−0.42	0.009 306

Notes: Solar System abundances are based mainly on element abundances in CI chondrite meteorites and in Sun's photosphere. CI chondrites are chosen because their volatile element contents are consistent with solar abundance ratios for several elements, and they have not undergone the fractionation processes experienced by other meteorites. Several compilations and estimates of solar abundances have been made over recent years. The above is one of the most recent and widely quoted.

Source: Lodders, K. (2003). Solar System abundances and condensation temperatures of the elements. *The Astrophysical Journal*, **591**, 1220–1247.

Table 1.2 Sun – physical properties

Radius	6.9595×10^8 m
Volume	1.4122×10^{27} m^3
Surface area	6.0865×10^{18} m^2
Mass	1.989×10^{30} kg
Mean density	1.409×10^3 kg m^{-3}
Surface gravity	274.0 m s^{-2}
Rotation rate (mean)	24.66 days
Absolute radiation	3.85×10^{26} W
Radiation at 1 ua	1367 W m^{-2}
Earth distance (mean) = 1 Astronomical Unit (ua)	$1.495\,979 \times 10^{11}$ m

Sources: Livingston, W. C. (2000). Sun. In *Allen's Astrophysical Quantities*, 4th edn, ed. A. N. Cox. New York: Springer-Verlag, pp. 339–380. Lodders, K. and Fegley, B. (1998). *The Planetary Scientist's Companion*. Oxford: Oxford University Press.

Table 1.3 Cosmic gas condensation sequence[a]

Temperature (K)	Mineral phase
1758 (1513)	Corundum Al_2O_3
1647 (1393)	Perovskite $CaTiO_3$
1625 (1450)	Melilite $Ca_2Al_2SiO_7$–$Ca_2MgSi_2O_7$
1513 (1362)	Spinel $MgAl_2O_4$
1471	Fe, Ni metal
1450	Diopside $CaMgSi_2O_6$
1444	Forsterite Mg_2SiO_4
1362	Anorthite $CaAl_2Si_2O_8$
1349	Enstatite $MgSiO_3$
<1000	Alkali feldspar $(Na,K)AlSi_3O_8$–$CaAl_2Si_2O_8$
<1000	Ferrous olivines, pyroxenes, $(Mg,Fe)_2SiO_4$, $(Mg,Fe)SiO_3$
700	Troilite FeS
405	Magnetite Fe_3O_4

Notes: a The sequence is calculated for a modelled cosmic gas at a pressure of 100 Pa. The temperature indicates the start of mineral formation, with the figure in parentheses indicating the temperature at which the mineral has subsequently reacted away.

Source: Wood, J. A. (1988). Chondritic meteorites and the solar nebula. *Annual Review of Earth and Planetary Sciences*, **16**, 53–72.

Table 1.4 The Solar System

Body	Mean distance from Sun (10^6 km)	Orbit (days)	Rotation (days)	Radius, equator (km)
Sun	–	–	24.66	695 508
Mercury	57.909	87.9694	58.6462	2 439.7
Venus	108.209	224.695	−243.0187	6 051.8
Earth	149.598	365.256	0.997 2697	6 378.14
Mars	227.937	686.980	1.025 96	3 397
Ceres[c]	413.715	1679.82	0.3781	471
Jupiter	778.412	4330.6	0.41354	71 492
Saturn	1 426.725	10 727.2	0.444 01	60 268
Uranus	2 870.972	30 717.7	−0.718 33	25 559
Neptune	4 498.253	60 215.9	0.671 252	24 764
Pluto[c]	5 906.376	90 803.7	−6.3872	1 195
Eris[c]	10 210[d]	20 345		∼600

Notes: *a* The Earth's dipole moment varies with time. Over the last 80 000 years it has varied between ∼2 and ∼11 × 10^{22} Am^2, see Laj *et al*. See also Table 1.5. *b* Mean temperature at the surface of planets. *c* Dwarf planet. (The International Astronomical Union (IAU) defined a category of 'dwarf planets' on 24 August 2006. Three bodies were initially included; others are likely to follow.) *d* Eris has a highly elliptical orbit.

Sources: Laj, C. *et al*. (1996). *Geophysical Research Letters*, **23**, 2045–2048. Lodders, K. and Fegley, Jr., B. (1998). *The Planetary Scientist's Companion*. Oxford: Oxford University Press. Zombeck, M. V. (2007). *Handbook of Space Astronomy and Astrophysics*. 3rd edn. Cambridge: Cambridge University Press. Tholen, D. J., Tejfel, V. G. and Cox, A. N. (2000). Planets and satellites. In *Allen's Astrophysical Quantities*, 4th edn, ed. A. N. Cox. New York: Springer-Verlag, pp. 293–313. Tholen, D. J. and Buie, M. W. (1997). *Icarus*, **125**, 245–260. Weaver, H. A. *et al*. (2006). *Nature*, **439**, 943–945.

Web pages: http://www.iau.org (September 2008); http://www.iau2006.org, http://antwrp.gsfc.nasa.gov (October 2006); http://nssdc.gsfc.nasa.gov/planetary/factsheet (September 2008).

Mass (10^{24} kg)	Density (kg m^{-3})	Gravity (m s^{-2})	Magnetic dipole moment[a] (A m^2)	Surface[b] T (K)	Satellites number
1989 000	1409	274.0			
0.3302	5430	3.701	3×10^{19}	440	0
4.8685	5243	8.870	$<8 \times 10^{17}$	730	0
5.9737	5515	9.820	8.02×10^{22}	290	1
0.6419	3940	3.727	1×10^{18}	183–268	2
0.000 95	2080	0.27		167	0
1898.7	1330	23.12	1.56×10^{27}	152	63
568.51	700	8.96	4.72×10^{25}	143	60
86.849	1300	8.69	3.83×10^{24}	68	27
102.44	1760	11.00	2.16×10^{24}	53	13
0.013	2030	0.81		58	3
~0.015				30	1

Table 1.5 Planets – physical properties in relation to Earth

Body	Volume	Mass	Gravity	Dipole moment[a]
Mercury	0.054	0.0553	0.38	0.0007
Venus	0.88	0.815	0.90	<0.0004
Earth	1.00 $(1.083\,21 \times 10^{21} \text{ m}^3)$	1.0000 $(5.9737 \times 10^{24} \text{ kg})$	1.00 (9.82 m s^{-2})	1[a] $(3.1 \times 10^{-5} \text{ T})$
Mars	0.149	0.1074	0.38	<0.0002
Ceres[b]	0.000 42	0.000 16	0.03	
Jupiter	1316	317.84	2.35	20 000
Saturn	755	95.169	0.91	600
Uranus	52	14.538	0.88	50
Neptune	44	17.148	1.12	25
Pluto[b]	0.005	0.0022	0.08	
Eris[b]	0.007	0.0025		

Notes: a These comparative data are for the average surface field strength along the planet's magnetic equator. See also Table 1.4 for data on the magnetic dipole moment. b dwarf planet.

Sources: Lodders, K. and Fegley, Jr., B. (1998). *The Planetary Scientist's Companion.* Oxford: Oxford University Press. Tholen, D. J., Tejfel, V. G. and Cox, A. N. (2000). Planets and satellites. In *Allen's Astrophysical Quantities*, 4th edn, ed. A. N. Cox. New York: Springer-Verlag, pp. 293–313. Van Allen, J. A. and Bagenal, F. (1999). Planetary magnetospheres and the interplanetary medium. In *The New Solar System*, 4th edn, ed. J. K. Beatty *et al.* Cambridge: Cambridge University Press.

Table 1.6 Planetary atmospheres: constituents

Planet	Main (volume %)											Minor
	H_2	He	CO_2	CO	CH_4	N_2	O_2	H_2O	Na	Ar	K	
Mercury	*	*					*		*		*	CO_2, N_2, Ne, Ar, Ca, Kr
Venus			96.5			3.5						H_2O, He, CO, Ne, SO_2, Ar
Earth			0.03			78.08	20.95	0–3		0.93		See Chapter 5
Mars			95.3	0.08		2.7	0.13			1.6		He, Ne, Ar, H_2O
Jupiter	86.4	13.6										HD, H_2O, CH_4, C_2H_6, NH_3, H_2S, Ne
Saturn	96.7	3.3										HD, CH_4, C_2H_6, NH_3
Uranus	82.5	15			2.3							HD
Neptune	80	19			1.5							HD, C_2H_6
Pluto					*	*						HD, C_2H_6

Notes: Data are not for available for all planets. An asterisk indicates presence, the available data are however very variable.

Sources: Atreya, S. K. et al. (1999). A comparison of the atmospheres of Jupiter and Saturn: deep atmospheric composition, cloud structure, vertical mixing, and origin. *Planetary and Space Science*, **47**, 1243–1262.

Bida T. A. et al. (2000). Discovery of calcium in Mercury's atmosphere. *Nature*, **404**, 159–161.

Lodders, K. and Fegley, Jr. B. (1998). *The Planetary Scientist's Companion*. Oxford: Oxford University Press.

Tholen, D. J. et al. (1999). Planets and satellites. In *Allen's Astrophysical Quantities*, 4th edn, ed. A. N. Cox. New York: Springer-Verlag.

1

Table 1.7 Composition of Mercury, Venus and Mars

Wt%	Mercury	Venus	Mars
Mantle and Crust			
SiO_2	47.1	52.9	45.39
Al_2O_3	6.4	3.8	2.89
MgO	33.7	37.6	29.71
CaO	5.2	3.6	2.35
Na_2O	0.08	1.6	0.98
K_2O	0.01	0.17	0.11
TiO_2	0.3	0.20	0.14
Cr_2O_3	3.3		0.68
MnO	0.06		0.37
FeO	3.7	0.24	17.22
P_2O_5			0.17
Total	99.85	100.11	100.01
Core[a]			
Fe	88.6	94.4	61.48
Co			0.38
Ni	5.5	5.6	7.67
Fe_3P			1.55
S	5.1		
FeS			28.97
Total	99.2	100.0	100.05

Notes: *a* The core of Mercury comprises 32.0%, of Venus 30.2% and of Mars 20.6% of the planet's total mass.

Sources: Mercury: Morgan, J. W. and Anders, E. (1980). Chemical composition of Earth, Venus and Mercury. *Proceedings of the National Academy of Sciences*, **77**, 6973–6977. Venus: Basalt Volcanism Study Project (1981). *Basaltic Volcanism on the Terrestrial Planets*. Pergamon Press. Mars: Lodders, K. and Fegley, B., Jr. (1997). An oxygen isotope model for the composition of Mars. *Icarus*, **126**, 373–394.

Table 1.8 Selected satellite names and properties[a]

Satellite	Name	Discovery date	Radius (km)	Mass (kg)
Earth				
	Moon		1737.4	7.3483×10^{22}
Mars				
I	Phobos	1877	$13.4 \times 11.2 \times 9.2$	1.063×10^{16}
II	Deimos	1877	$7.5 \times 6.1 \times 5.2$	2.38×10^{15}
Jupiter				
I	Io	1610	$1830 \times 1819 \times 1815$	8.9316×10^{22}
II	Europa	1610	1565	$4.799\ 82 \times 10^{22}$
III	Ganymede	1610	2634	$1.481\ 86 \times 10^{23}$
IV	Callisto	1610	2403	$1.075\ 93 \times 10^{23}$
V	Amalthea	1892	$131 \times 73 \times 67$	7.2×10^{18}
VI	Himalia	1904	85	9.5×10^{18}
VII	Elara	1905	40	8×10^{17}
VIII	Pasiphae	1908	18	2×10^{17}
IX	Sinope	1914	14	8×10^{16}
X	Lysithea	1938	12	8×10^{16}
XI	Carme	1938	15	9×10^{16}
XII	Ananke	1951	10	4×10^{16}
XIII	Leda	1974	5	6×10^{15}
XIV	Thebe	1980	55×45	8×10^{17}
XV	Adrastea	1979	$13 \times 10 \times 8$	2×10^{16}
XVI	Metis	1980	20×20	9×10^{16}
Saturn				
I	Mimas	1789	$209 \times 196 \times 191$	3.75×10^{19}
II	Enceladus	1789	$256 \times 247 \times 245$	7×10^{19}
III	Tethys	1684	$536 \times 528 \times 526$	6.27×10^{20}
IV	Dione	1684	560	1.10×10^{21}
V	Rhea	1672	764	2.31×10^{21}
VI	Titan	1655	2 575	1.3455×10^{23}
VII	Hyperion	1848	$180 \times 140 \times 112.5$	2×10^{19}
VIII	Iapetus	1671	718	1.6×10^{21}

(*cont.*)

Table 1.8 (*cont.*)

Satellite	Name	Discovery date	Radius (km)	Mass (kg)
IX	Phoebe	1898	110	4×10^{17}
X	Janus	1966	$97 \times 95 \times 77$	1.92×10^{18}
XI	Epimetheus	1966/1978	$69 \times 55 \times 55$	5.4×10^{17}
XII	Helene	1980	$18 \times 16 \times 15$	
XIII	Telesto	1980	$15 \times 12.5 \times 7.5$	
XIV	Calypso	1980	$15 \times 8 \times 8$	
XV	Atlas	1980	$18.5 \times 17.2 \times 13.5$	
XVI	Prometheus	1980	$74 \times 50 \times 34$	
XVII	Pandora	1980	$55 \times 44 \times 31$	
XVIII	Pan	1990	10	
Uranus				
I	Ariel	1851	$581 \times 578 \times 578$	1.35×10^{21}
II	Umbriel	1851	584.7	1.17×10^{21}
III	Titania	1787	788.9	3.53×10^{21}
IV	Oberon	1787	761.4	3.01×10^{21}
V	Miranda	1948	$240 \times 234 \times 233$	6.6×10^{19}
VI	Cordelia	1986	13	
VII	Ophelia	1986	15	
VIII	Bianca	1986	21	
IX	Cressida	1986	31	
X	Desdemona	1986	27	
XI	Juliet	1986	42	
XII	Portia	1986	54	
XIII	Rosalind	1986	27	
XIV	Belinda	1986	33	
XV	Puck	1986	77	
XVI	Caliban	1997	30	
XVII	Sycorax	1997	60	
Neptune				
I	Triton	1846	1352.6	2.14×10^{22}
II	Nereid	1949	170	2×10^{19}

Table 1.8 (*cont.*)

Satellite	Name	Discovery date	Radius (km)	Mass (kg)
III	Naiad	1989	29	
IV	Thalassa	1989	40	
V	Despina	1989	74	
VI	Galatea	1989	79	
VII	Larissa	1982	104×89	
VIII	Proteus	1989	$218 \times 208 \times 201$	
Pluto[b]				
I	Charon	1978	593	1.62×10^{21}
II	Nix	2005	~23	
III	Hydra	2005	~30	
Eris[b]				
I	Dysnomia	2005	~170	

Notes: *a* New technology is allowing the discovery of several new satellites. At the IAU General Assembly in Sydney, July 2004, names were given to 35 satellites (of Jupiter, Saturn and Uranus) – see web page (which includes a brief discussion on naming): http://www.iau.org/SATELLITES_OF_PLANETS.248.0.html?&0 (October 2006). *b* dwarf planet.

Sources: Tholen, D. J. *et al.* (2000). Planets and satellites. In *Allen's Astrophysical Quantities*, 4th edn, ed. A. N. Cox. New York: Springer-Verlag, pp. 293–313. Web page (October 2006): http://www.iau2006.org.

Table 1.9 Earth's orbital variations

Parameter	Main periodicities (ka)	Resultant periodicities (ka)	Main climatic effect
Precession (as combined effect of orbital eccentricity and axial precession)	17, 19, 22, 24	21.7	Change to relative length and duration of seasons
Obliquity (tilt of axis) varying between 21.8° and 24.4°	41		Strength of seasonal variation
Eccentricity (orbit shape)	95, 125, ~400	~100, ~400	Slight seasonal variation

Note: The above table gives only the main modes of the orbital variations and is, therefore, a simplification. M. Milankovitch (1879–1958) used the variations within the three orbital parameters to model the changes in insolation with time. The results are sometimes referred to as 'Milankovitch cycles'.

Sources: Hinnov, L. A. (2004). Earth's orbital parameters and cycle stratigraphy. In *A Geologic Time Scale (2004)*, ed. Gradstein, F. *et al.* Cambridge: Cambridge University Press. Weedon, G. (2003). *Time-series Analysis and Cyclostratigraphy. Examining Stratigraphic Records of Environmental Cycles.* Cambridge: Cambridge University Press.

Table 1.10a The Moon – physical properties

Mean distance from Earth	384 401 km
Mean radius	1738.3 km
Mass	7.353×10^{22} kg
Volume	2.200×10^{10} km^3
Surface area	3.792×10^8 km^2
Surface temperature	120–390 K
Heat flow (average)	$\sim$29 mW m^{-2}
Surface gravity	1.624 m s^{-2}
Bulk density	3.341 kg m^{-3}
Sidereal period	27.321 661 days
Synodic month	29.530 588 days
Orbital eccentricity	0.0549
Mean orbital velocity	1.023 km s^{-1}

Sources: Heiken, G. H., Vaniman, D. T. and French, B. M., eds (1991). *Lunar Sourcebook, a User's Guide to the Moon.* Cambridge: Cambridge University Press. Lodders, K. and Fegley, Jr., B. (1998). *The Planetary Scientist's Companion.* Oxford: Oxford University Press. Tholen, D. J. *et al.* (2000). Planets and satellites. *Allen's Astrophysical Quantities*, 4th edn, ed. A. N. Cox. New York: Springer-Verlag.

Table 1.10b The Moon – internal structure

Unit	Depth, km
Crust[a]	0–50 average
Upper mantle[b]	50–560
Middle mantle	~560–1150
Lower mantle	~1150–1400
Outer core, fluid	~1400–1580
Inner core, solid[c]	~1580–centre

Notes: a Crustal thickness varies significantly, especially from nearside to farside.

b A seismic discontinuity exists at ~560 km depth.

c Existence of a solid core has not been confirmed.

Source: Based mainly on Wieczorek, M. A. *et al.* (2006). The constitution and structure of the lunar interior. In *New Views of the Moon*, ed. L. J. Bradley *et al.*, *Reviews in Mineralogy & Geochemistry*, vol. 60. Mineralogical Society of America and Geochemical Society.

Table 1.11 The Moon – bulk chemical composition[a]

Major elements (wt%)

8	O	44.11
12	Mg	19.3
13	Al	3.17
14	Si	20.3
20	Ca	3.22
26	Fe	10.6

Minor/trace elements (μg/g)

3	Li	0.83	57	La	0.90
4	Be	0.81	58	Ce	2.34
5	B	0.54	59	Pr	0.34
11	Na	600	60	Nd	1.74
15	P	573	62	Sm	0.57
19	K	83	63	Eu	0.21
21	Sc	19	64	Gd	0.75
22	Ti	1800	65	Tb	0.14
23	V	150	66	Dy	0.93
24	Cr	4200	67	Ho	0.21
25	Mn	1200	68	Er	0.61
37	Rb	0.28	69	Tm	0.088
38	Sr	30	70	Yb	0.61
39	Y	5.1	71	Lu	0.093
40	Zr	14	72	Hf	0.42
41	Nb	1.1	74	W (ng/g)	740
55	Cs (ng/g)	12	90	Th (ng/g)	125
56	Ba	8.8	92	U (ng/g)	33

Note: a The composition of the bulk moon is a controversial topic, see for example, Warren, P. H. (2004). The Moon. In *Meteorites, Comets and Planets*, ed. A. M. Davis, vol. 1, *Treatise on Geochemistry*, ed. H. D. Holland and K. K. Turekian. Oxford: Elsevier- Pergamon, pp. 559–599.

Source: Taylor, S. R. (1982). *Physics of Earth and Planetary Interiors*, **29**, 233 241.

Table 1.12 Lunar stratigraphy[a]

Period	Epoch	Age at base (Ga)	Events during period
Copernican		1.10	Little volcanism Formation of Tycho and Copernicus craters
Erastothenian		3.20	Basalt flows
Imbrium	Late	3.80	Cessation of enormous impacts
	Early	3.85	Formation of Imbrium and Orientale basins
Nectarian		~3.92	Continued heavy bombardment Formation of Nectaris, Serenitatis and Crisium basins
Pre-Nectarian		~4.2	End of main differentiation Intense impact bombardment
		~4.55	Moon formation and differentiation

Note: *a* Lunar stratigraphy is not well resolved. Several schemes have been proposed. This one is based on Wilhems, D. E. (1987). *The Geologic History of the Moon.* U.S. Geol. Survey Prof. Paper 1348. For a brief review on different stratigraphic schemes see Hiesinger, H. and Head III, J. W. (2006). *New Views of the Moon*, ed. L. J. Bradley *et al.*, *Reviews in Mineralogy & Geochemistry*, vol. 60. Mineralogical Society of America and Geochemical Society.

Table 1.13 Composition of lunar rock types – selected examples

| Rock type | Mare basalts | | | | KREEP | Anorthosite | Breccia | Soils | |
	High-Ti A	Low-Ti B	Al-rich C	Very-low-Ti D	E	F	G	Mare H	Highland I
wt %									
SiO_2	39.8	45.3	46.4	46.0	50.8	44.5	45.1	46.3	45.0
TiO_2	10.5	4.7	2.6	1.1	2.06	0.016	0.32	3.0	0.54
Al_2O_3	10.4	10.0	13.6	12.1	15.0	35.6	30.4	12.9	27.3
Cr_2O_3	0.25	0.31	–	0.30	0.33	–	0.06	0.34	0.33
FeO	19.8	20.2	16.8	22.1	10.4	0.21	3.34	15.1	5.1
MnO	0.30	0.28	0.26	0.28	0.16	–	0.05	0.22	0.30
MgO	6.7	7.0	8.5	6.0	9.34	0.26	3.14	9.3	5.7
CaO	11.1	11.4	11.2	11.6	9.60	20.4	17.0	10.7	15.7
Na_2O	0.40	0.29	–	0.26	0.77	0.364	0.48	0.54	0.46
K_2O	0.06	0.06	0.10	0.02	0.61	0.017	0.04	0.31	0.17
Total	99.31	99.54	99.00	100.00	99.07	101.367	99.90	98.71	100.6

(*cont.*)

1

Table 1.13 (*cont.*)

Rock type	Mare basalts				KREEP	Anorthosite	Breccia	Soils	
	High-Ti A	Low-Ti B	Al-rich C	Very-low-Ti D	E	F	G	Mare H	Highland I
$\mu g/g$									
P	–	–	–	–	3054	40	–	1746	480
Sc	–	58	55	57	22	0.42	–	40.2	8.0
V	63	102	–	140	60	–	–	110	20
Ni	–	5.9	14	30	12.5	9	–	–	–
Rb	0.62	0.91	2.19	–	16.2	0.17	–	–	–
Sr	161	148	98.0	110.0	187	188	–	140	170
Ba	108	73.6	146.0	50.0	744	6.2	–	430	130
La	15.5	6.53	13.0	2.9	70.8	0.15	–	35.6	10.8
Ce	47.2	19.2	34.5	8.6	179	0.33	–	85	28
Nd	40.0	15.4	21.9	7.0	114	0.19	–	57	19
Sm	14.4	5.68	6.56	2.1	33	0.053	–	17.3	4.79
Eu	1.81	1.23	1.21	0.83	2.56	0.81	–	1.85	1.05
Gd	19.5	7.89	8.59	–	45.4	0.056	–	–	–
Tb	–	–	–	0.45	5.3	0.0085	–	3.7	1.0
Dy	12.9	9.05	10.5	2.9	–	0.054	–	22	6.0

	A	B	C	D	E	F	G	H
Ho	–	–	0.71	–	–	–	5.0	1.4
Er	13.6	5.57	6.51	–	–	0.034	–	–
Tm	–	–	–	–	–	–	1.8	0.55
Yb	13.2	5.46	6	2.0	21.3	0.034	13.0	3.40
Lu	1.0	–	–	0.31	2.84	0.0043	1.85	0.49
Hf	–	3.1	9.8	1.4	23.6	0.014	11.8	3.30
Zr	309	128	215	50	–	–	–	–
Th	1.1	1.0	2.1	0.20	10.0	0.004	5.40	1.85
U	–	0.26	0.59	–	2.80	0.002	–	0.4

Notes: Lunar samples are: A – a high-Ti mare basalt (10003) from the Apollo 11 site; B – a low-Ti mare basalt (12051) from the Apollo 12 site; C – a low-Ti, aluminous mare basalt (14053) from the Apollo 14 site; D – a very-low-Ti mare basalt (24174,7) from the Luna 24 site; E – a KREEP basalt (15386) from the Apollo 15 site. KREEP are rocks with component enriched in potassium, rare earths and phosphorus; F – an anorthosite (15415) from the Apollo 15 site; G – the bulk average composition of matrix samples of fragmental breccias; H – average composition (major elements) of Apollo 12 soils, trace element composition for one sample (12001,599); I – average composition (major elements) Apollo 16 soils, trace element composition for one sample (64501,122).

Sources: McKay, D. S. et al. (1991). The lunar regolith. In Lunar Sourcebook. A User's Guide to the Moon, ed. G. Heiken et al. Cambridge: Cambridge University Press. Taylor, G. J. et al. (1991). Lunar rocks. In Lunar Sourcebook. A User's Guide to the Moon, ed. G. Heiken et al. Cambridge: Cambridge University Press.

Table 1.14 Meteorite classification and numbers

Class[a]	Group		Number
Stones: chondrites			
Ordinary			
	H (high-Fe[b])		6962
	L (low-Fe)		6213
	LL (low-Fe, low-metal)		1048
	Other		42
Carbonaceous	C		561
Enstatite			
	E		38
	EH (high-Fe)		125
	EL (low-Fe)		38
Other chondrites			163
		Total chondrites	15 190
Stones: achondrites			
Aubrites			46
Ureilites			92
Diogenites	HED Group		94
Eucrites			200
Howardites			93
Lunar			18
Martian (SNC[c])			15
Other classes and unclassified			52
		Total achondrites	610
Stones: ungrouped and unclassified			5714
		Total stones	21 514
Stony-irons			
Mesosiderites			66
Pallasites			50
		Total stony-irons	116

Table 1.14 (*cont.*)

Class^a	Group	Number
Irons		
	IAB	131
	IIAB	103
	IID	16
	IIE	18
	IIIAB	230
	IIICD	41
	IVA	64
	Other classes, ungrouped and unclassified	262
	Total irons	865
	Total meteorites (inc. 12 unknowns)	22 507

Notes: *a* The classes given are ones containing 15 or more meteorites as recorded in 2000. Current data on numbers of meteorites in different classes can be found on the Meteoritical Society web site: http://tin.er.usgs.gov/meteor/metbull.php. *b* Fe = concentration of iron. *c* This class is sometimes referred to as 'SNC' after the names of its three principal meteorites: Shergotty, Nakhla and Chassigny. Classification of stones based on chemical composition and petrological texture; of stony irons on mineralogy (mesosiderites contain metal, olivine and other silicates; pallasites contain mainly metal and olivine); and of irons on chemical composition. See Tables 1.15b and d for further information on classification of carbonaceous chondrites and iron meteorites. Individual meteorites are named after their place of find or fall – usually taken from the nearest inhabited place. For uninhabited areas (deserts etc.) a system is used that combines geographical and date information.

Sources: Grady, M. M. (2000). *Catalogue of Meteorites*, 5th edn. Cambridge: Cambridge University Press. Hutchison, R. (2004). *Meteorites. A Petrologic, Chemical and Isotopic Synthesis*. Cambridge: Cambridge University Press.

Table 1.15a Meteorite compositions: CI chondrites

Element abundances, μg/g except where stated

1	H wt%	2.1	33	As	1.73
2	He	0.009 17	34	Se	19.7
3	Li	1.46	35	Br	3.43
4	Be	0.0252	36	Kr	5.22×10^{-5}
5	B	0.713	37	Rb	2.13
6	C wt%	3.518	38	Sr	7.74
7	N	2940	39	Y	1.53
8	O wt%	45.8200	40	Zr	3.96
9	F	60.6	41	Nb	0.265
10	Ne	0.000 18	42	Mo	1.02
11	Na	5010	44	Ru	0.692
12	Mg wt%	9.5870	45	Rh	0.141
13	Al	8500	46	Pd	0.588
14	Si wt%	10.6500	47	Ag	0.201
15	P	920	48	Cd	0.675
16	S wt%	5.4100	49	In	0.0788
17	Cl	704	50	Sn	1.68
18	Ar	0.001 33	51	Sb	0.152
19	K	530	52	Te	2.33
20	Ca	9070	53	I	0.48
21	Sc	5.83	54	Xe	1.74
22	Ti	440	55	Cs	0.185
23	V	55.7	56	Ba	2.31
24	Cr	2590	57	La	0.232
25	Mn	1910	58	Ce	0.621
26	Fe wt%	18.2800	59	Pr	0.0928
27	Co	502	60	Nd	0.457
28	Ni wt%	1.0640	62	Sm	0.145
29	Cu	127	63	Eu	0.0546
30	Zn	310	64	Gd	0.198
31	Ga	9.51	65	Tb	0.0356
32	Ge	33.2	66	Dy	0.238

Table 1.15a (*cont.*)

Element abundances, μg/g except where stated

67	Ho	0.0562	77	Ir	0.470
68	Er	0.162	78	Pt	1.004
69	Tm	0.0237	79	Au	0.146
70	Yb	0.163	80	Hg	0.314
71	Lu	0.0237	81	Tl	0.143
72	Hf	0.115	82	Pb	2.56
73	Ta	0.0144	83	Bi	0.110
74	W	0.089	90	Th	0.0309
75	Re	0.037	91	U	0.0084
76	Os	0.486			

Notes: These data are based on the mean of element abundances in the five CI chondrite falls. CI chondrite composition is used for normalising other geochemical data – see Table 1.16. CI chondrites are used as a reference for abundances because their volatile element contents are consistent with solar abundance ratios for several elements, and they have not undergone the fractionation processes experienced by other meteorites.

Source: Lodders, K. (2003). Solar System abundances and condensation temperatures of the elements. *The Astrophysical Journal*, **591**, 1220–1247.

Table 1.15b Meteorite compositions: carbonaceous chondrites (selected elements)[a]

Z	Element	Unit	Chondrite group					
			CM	CV	CO	CK	CR	CH
11	Na	wt%	0.41	0.33	0.41	0.319	0.323	0.180
12	Mg	wt%	11.7	14.5	14.5	14.8	13.9	11.3
13	Al	wt%	1.18	1.75	1.43	1.61	1.27	1.05
14	Si	wt%	12.9	15.6	15.9	15.1	15.3	13.5
16	S	wt%	3.3	2.2	2.0	1.7	1.9	0.35
19	K	µg/g	400	310	345	285	303	200
20	Ca	wt%	1.27	1.90	1.58	1.72	1.38	1.3
21	Sc	µg/g	8.2	11.4	9.6	11.0	8.49	7.5
22	Ti	µg/g	580	980	780	940	540	650
23	V	µg/g	75	96	92	96	79	63
24	Cr	µg/g	3050	3600	3550	3660	3750	3100
25	Mn	µg/g	1700	1450	1650	1460	1700	1020
26	Fe	wt%	21.0	23.5	24.8	23.6	24.0	38.0
27	Co	µg/g	575	655	688	637	667	1100
28	Ni	wt%	1.20	1.34	1.40	1.27	1.36	2.57
37	Rb	µg/g	1.7	1.25	1.45		1.1	
38	Sr	µg/g	10.1	15.3	12.7	15	10	
39	Y	µg/g	2.0	2.4	2.4	2.7		
42	Mo	µg/g	1.5	2.1	1.9	3.8	1.4	2.0
57	La	ng/g	317	486	387	462	342	290
58	Ce	ng/g	838	1290	1020	1.27	0.75	870
60	Nd	ng/g	631	990	772	0.99	0.79	
62	Sm	ng/g	200	295	240	284	210	185
63	Eu	ng/g	76	113	94	108	84	76
64	Gd	ng/g	276	415	337	440	320	290
65	Tb	ng/g	47	65	57		50	50
66	Dy	ng/g	330	475	404	490	280	310
67	Ho	ng/g	77	110	94	100	100	70
70	Yb	ng/g	222	322	270	311	236	210
71	Lu	ng/g	33	48	40	44	35	30

Table 1.15b (*cont.*)

Z	Element	Unit	Chondrite group					
			CM	CV	CO	CK	CR	CH
72	Hf	ng/g	186	194	178	250	150	140
74	W	ng/g	140	190	160	180	110	150
75	Re	ng/g	46	65	55	60	50	73
76	Os	ng/g	640	825	790	813	679	1150
77	Ir	ng/g	595	760	735	767	642	1070
78	Pt	ng/g	1100	1250	1200	1300	980	1700
79	Au	ng/g	165	144	184	136	139	250
92	U	ng/g	11	17	13	15	13	

Notes: *a* Data are mostly mean values. Each carbonaceous chondrite group symbol is assigned after the name of one of their representative members e.g. the Murchison meteorite in the CM group. See Table 1.15a for CI chondrite data.

Sources: Lodders, K. and Fegley, B., Jr. (1998). *The Planetary Scientist's Companion*. Oxford: Oxford University Press. Hutchison, R. (2004). *Meteorites. A Petrologic, Chemical and Isotopic Synthesis*. Cambridge: Cambridge University Press. Kalleymeyn, G. W. *et al*. (1991). The compositional classification of chondrites: V. The Karoonda (CK) group of carbonaceous chondrites. *Geochimica et Cosmochimica Acta*, **55**, 881–892. Kalleymeyn, G. W. *et al*. (1994). The compositional classification of chondrites: VI. The CR carbonaceous chondrite group. *Geochimica et Cosmochimica Acta*, **58**, 2873–2888. Wasson J. T. and Kalleymeyn, G. W. (1988). Compositions of chondrites. *Philosophical Transactions of the Royal Society, London*, A **325**, 535–544.

Table 1.15c Meteorite compositions: ordinary and enstatite chondrites

	Ordinary			Enstatite	
	H	L	LL	EL	EH
Si	16.9	18.5	18.9	18.6	16.7
Ti	0.060	0.063	0.062	0.058	0.045
Al	1.13	1.22	1.19	1.05	0.81
Cr	0.366	0.388	0.374	0.305	0.315
Fe	27.5	21.5	18.5	22.0	29.0
Mn	0.232	0.257	0.262	0.163	0.220
Mg	14.0	14.9	15.3	14.1	10.6
Ca	1.25	1.31	1.3	1.01	0.85
Na	0.64	0.70	0.70	0.580	0.680
K	0.078	0.083	0.079	0.074	0.080
P	0.108	0.095	0.085	0.117	0.200
Ni	1.60	1.20	1.02	1.30	1.75
Co	0.081	0.059	0.049	0.067	0.084
S	2.0	2.2	2.3	3.3	5.8
C	0.11	0.09	0.12	0.36	0.40
O	35.7	37.7	40.0	31.0	28.0
Total	101.755	100.265	100.241	94.084	95.534

Note: The above are averages in weight %.

Source: Hutchison, R. (2004). *Meteorites. A Petrologic, Chemical and Isotopic Synthesis*. Cambridge: Cambridge University Press.

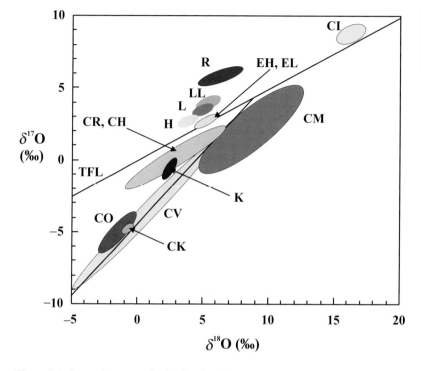

Figure 1.1 Oxygen isotope ratio plot for chondrite meteorites. The delta values of the isotopic ratios $^{17}O/^{16}O$ and $^{18}O/^{16}O$, measured in relation to SMOW (oxygen isotope value of Standard Mean Ocean Water), are used to help classify these meteorites (see Chapter 8, and Tables 8.15 and 8.18, for further information and standard notation on stable isotope data). The letters denote the chondrite group names, see Tables 1.14, 1.15a and b. R is for the R group (containing just one meteorite) – chemically similar to the H group. K is for the Kakangari triplet – a group of just three meteorites resembling chondrites with respect to major element composition but more highly reduced. The line marked TFL is the 'terrestrial fractionation line', which defines the isotopic ratios of virtually all Earth materials. The line drawn through the CV field defines the trend of anhydrous, high-temperature chondrules and calcium–aluminium-rich inclusions in CV chondrites.

Source: R. Hutchison, (2004). *Meteorites. A Petrologic, Chemical and Isotopic Synthesis.* Cambridge University Press, with permission.

Table 1.15d Meteorite compositions: irons

Class:	IAB	IC	IIAB	IIC	IID	IIE
Range[a]						
Ni wt%	6.5–60.8	6.1–6.8	5.3–6.4	9.3–11.5	9.6–11.3	7.5–9.7
Ga (μg/g)	2–100	49–55	46–62	37–39	70–83	21–28
Ge (μg/g)	2–520	212–247	107–185	88–114	82–98	62–75
Ir (μg/g)	0.02–6	0.07–2.1	0.01–0.9	4–11	3.5–18	1–8
Average						
Ni wt%	6.4	7.1	6.15	10.8	11.2	9.13
Ga (μg/g)	96	51.7	58.3	37.3	76.2	24.4
Ge (μg/g)	400	230	173	94.6	87	68.3
Ir (μg/g)	2.7	0.38	10	6.4	9.9	4.1
P (mg/g)	2.1	4.3	6	5.3	9.8	
Cr (μg/g)	16	70	38	87	31	
Co (mg/g)	4.6	4.6	5.3	6.5	4.7	4.7
Cu (μg/g)	132	160	129	260	280	416
Mo (μg/g)	8.2	7.7	6.9	8.4	9.4	6.8
W (μg/g)	1.6	1.3	2.1		2.4	0.78
Pd (μg/g)	3.5	3.5	2.6	6	5.3	5.3
Au (μg/g)	1.5	1.14	1.1	1.1	1.1	1.6
As (μg/g)	11	11	9.9	8.2	10	16
Sb (ng/g)	270	98	201	150	220	300

Note: a Iron meteorite classification has evolved as more chemical data became available. It is based primarily on the contents of Ni, Ga and Ge. Classes IAB and IIICD are, however, overlapping and it is not possible to give an average composition for IIICD separately.

Sources: Hutchison, R. (2004). *Meteorites. A Petrologic, Chemical and Isotopic Synthesis*. Cambridge: Cambridge University Press. Mittlefehldt, D. W. *et al.* (1998). Non-chondritic meteorites from asteroidal bodies. In *Planetary Materials*, ed. J. J. Papike, *Reviews in Mineralogy*, vol. 36. Washington, DC: Mineralogical Society of America.

IIF	IIIAB	IIIE	IIIF	IVA	IVB
10.6–14.3	7.1–10.5	8.2–9.0	6.8–8.5	7.4–9.4	16.0–18.0
8.9–11.6	16–23	17–19	6.3–7.3	1.6–2.4	0.17–0.27
99–193	27–47	34–37	0.7–1.1	0.09–0.14	0.003–0.07
0.75–23	0.01–20	0.01–6	0.006–7.9	0.4–4	13–38
12.9	8.49	8.49	7.95	8	17.1
9.8	19.7	17.6	6.82	2.15	0.221
140	38.9	35.6	0.91	0.124	0.0526
6.2	4.1	4.1	3.2	1.9	22
2.6	5.6	5.6	2.2	0.88	1.0
	40	40	210	140	87.7
7.0	5.0	5.0	3.6	4.0	7.4
300	160	160	170	150	12
	7.2	7.2	7.2	5.9	27
1.0	1.0	1.0	1.2	0.6	3.0
	3.5	3.5	4.4	4.6	12
1.5	1.2	1.2	0.91	1.5	0.15
16	10.5	10.5	11	7.6	1.1
250	265	265	86	9	1.5

Table 1.16 Chondrite normalising data: rare-earth elements

	Data set				
Element	Wakita	Haskin	Evensen	Anders	Sun-McD.
La	0.340	0.330	0.244 6	0.2347	0.237
Ce	0.910	0.880	0.637 9	0.6032	0.612
Pr	0.121	0.112	0.096 37	0.0891	0.095
Nd	0.640	0.600	0.473 8	0.4524	0.467
Sm	0.195	0.181	0.154 0	0.1471	0.153
Eu	0.073	0.069	0.058 02	0.0560	0.058
Gd	0.260	0.249	0.204 30	0.1966	0.2055
Tb	0.047	0.047	0.037 45	0.0363	0.0374
Dy	0.300		0.254 1	0.2427	0.2540
Ho	0.078	0.070	0.056 70	0.0556	0.0566
Er	0.200	0.200	0.166 0	0.1589	0.1655
Tm	0.032	0.030	0.025 61	0.0242	0.0255
Yb	0.220	0.200	0.165 1	0.1625	0.170
Lu	0.034	0.034	0.025 39	0.0243	0.0254

Notes: Chondritic meteorite compositions are often used as reference against which elemental compositions of other materials may be normalised for purposes of comparison and diagramatic presentation. Several data sets are in use for normalising rare-earth element data; those above are some of the most common ones. See also Table 1.15a. Data are not always absolute – some have been manipulated, for example the 'Evensen' data set were derived from relative abundance data multiplied by 0.2446 to produce a best fit to absolute abundances measured on C1 chondrites.

Wakita: a composite of 12 chondrites. Wakita, H. *et al.* (1971). Abundances of the 14 rare-earth elements and 12 other trace-elements in Apollo 12 samples: five igneous and one breccia rocks and four soils. *Proceedings of the Second Lunar Science Conference*, vol. 2. Cambridge: The MIT Press, pp. 1319–1329.

Haskin: Haskin, L. A. *et al.* (1968). Relative and absolute terrestrial abundances of the rare earths. In *Origin and Distribution of the Elements*, vol. 1, ed. L. H. Ahrens. Oxford: Pergamon Press, pp. 889–911.

Evensen: Evensen, N. M. *et al.* (1978). Rare earth elements in chondritic meteorites. *Geochimica et Cosmochimica Acta*, **42**, 1199–1212.

Anders: Anders, E. and Grevesse, N. (1989). Abundances of the elements: meteoritic and solar. *Geochimica et Cosmochimica Acta*, **43**, 197–214.

Sun-McD: Sun, S.-S. and McDonough, W. F. (1989). Chemical and isotopic systematics of oceanic basalts: implications for mantle composition and processes. In *Magmatism in the Ocean Basins*, ed A. D. Saunders and M. J. Norry, *Geological Society Special Publication*, vol. 42. Oxford: Blackwell Scientific Publications.

2 Solid Earth

Table 2.1 Earth stucture and geotherm

Layer	Depth range (km)	Mass (10^{21} kg)	Mass fraction	Average density (kg m^{-3})	Temperature estimates (K)
Oceanic crust[a]	0–6	6	0.1	3 000	0–~1300
Continental crust	0–30	19	0.3	2 700	0–~1300
Upper mantle	(6 or 30)–410	615	10.3	3 350	1273–1622
Transition zone	410–660	415	7.0	3 860	1600
Lower mantle	660–2886	2955	49.6	4 870	1830–3000
Outer core	2886–5140	1867	31.1	11 000	3573–3800 (CMB)
					4000–5800 (ICB)
Inner core	5140–6371	98	1.6	12 950	4676–6600
Whole Earth		5975		5 515	

Note: *a* It is estimated that there is an average sediment thickness of 0.5 km on the floor of the oceans. CMB = core–mantle boundary, ICB = inner-core boundary

Sources: Schubert, G., Turcotte, D. L. and Olsen, P. (2001). *Mantle Convection in the Earth and Planets*. Cambridge: Cambridge University Press. Geotherm based mainly on data in Poirier, J.-P. (2000). *Introduction to the Physics of the Earth's Interior*, 2nd edn. Cambridge: Cambridge University Press.

Table 2.2 Earth properties

Equatorial radius	6378.1 km
Polar radius	6356.8 km
Volume	1.0832×10^{21} m^3
Volume of core	1.77×10^{20} m^3
Volume of mantle	9.06×10^{20} m^3
Total area	5.10×10^{14} m^2
Land area	1.48×10^{14} m^2
Continental area including margins	2.0×10^{14} m^2
Total land area%	29%
Water area	3.62×10^{14} m^2
Water area%	71%
Ocean area excluding continental margins	3.1×10^{14} m^2
Mean land elevation	875 m
Mean ocean depth	3795 m
Highest land point	8848 m
Deepest ocean point	$-11\,040$ m
Surface heat flow (mean)	87 mW m^{-2}
Total geothermal flux	44.3 TW
Mean geothermal flux	0.08 W m^{-2}
Continental heat flow (mean)	65 mW m^{-2}
Oceanic heat flow (mean)	101 mW m^{-2}

Sources: Turcotte, D. L. and Schubert, G. (2002) *Geodynamics*, 2nd edn. Cambridge: Cambridge University Press. Stacey, F. D. (1992). *Physics of the Earth*, 3rd edn. Brisbane: Brookfield Press. Yoder, C. F. (1995). Astrometric and geodetic properties of Earth and the Solar System. In *Global Earth Physics. A Handbook of Physical Constants*, ed. T. J. Ahrens. Washington, DC: American Geophysical Union.

Table 2.3 Main tectonic plates: areas and velocities

Plate	Area (10^6 km^2)	Continental area (10^6 km^2)	Average velocity (cm.a^{-1})
N. American	60	36	1.1
S. American	41	20	1.3
Pacific	108	0	8.0
Antarctic	59	15	1.7
Indian	60	15	6.1
African	79	31	2.1
Eurasian	69	51	0.7
Nazca	15	0	7.6
Cocos	2.9	0	8.6
Caribbean	3.8	0	2.4
Philippine	5.4	0	6.4
Arabian	4.9	4.4	4.2

Sources: Forsyth, D. and Uyeda, S. (1975). On the relative importance of the driving forces of plate motion. *Geophysical Journal of the Royal Astronomical Society*, **43**, 163–200. Schubert, G. *et al.* (2001). *Mantle Convection in the Earth and Planets*. Cambridge: Cambridge University Press.

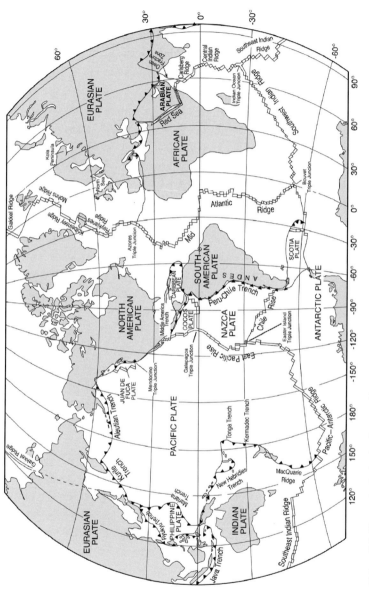

Figure 2.1 The major tectonic plates, mid-ocean ridges, trenches and transform faults. **Source:** Fowler, C. M. R. (2005). *The Solid Earth. An Introduction to Global Geophysics.* Cambridge: Cambridge University Press, with permission.

2

Table 2.4 Continents and oceans

Name	Area (km²)	Lowest point below sea level		Highest elevation	
		Name	Height (m)	Name	Height (m)
Continents					
Africa	30 293 000	Lake Assal, Djibouti	−156	Mt Kilimanjaro, Tanzania	5895
Antarctica	13 975 000	Bently subglacial trench	−2555	Vinson Massif	4897
Asia	44 493 000	Dead Sea	−400	Mt Everest, China/Nepal	8848
Australia + Oceania	8 945 000	Lake Eyre, Australia	−15	Puncak Jaya, Indonesia	5030
Europe	10 245 000	Caspian Sea, SW Asia	−29	Mt El'brus, Russia	5642
North America	24 454 000	Death Valley, California, USA	−86	Mt McKinley, Alaska, USA	6194
South America	17 838 000	Peninsula Valdes, Argentina	−40	Mt Aconcagua, Argentina	6960
		Mean depth (m)		**Greatest depth (m)**	
Oceans					
Arctic	9 485 000	1330		Molloy Deep	5 680
Atlantic	86 557 000	3700		Puerto Rica Trench	8 648
Indian	73 427 000	3900		Java Trench	7 725
Pacific	166 241 000	4300		Mariana Trench	11 040
Southern	20 327 000	4500		South Sandwich Trench	7 235

Notes: Areas of oceans depend on the chosen boundaries, about which there is no standard agreement. The areas used here are those given in Pilson (1998).

Sources: Central Intelligence Agency (2008). *The World Factbook.* https://www.cia.gov/library/publications/the-world-factbook/index.html accessed 4/2008. Marsden, H., ed. (2007). *Chambers Book of Facts.* Edinburgh: Chambers Harrap Publishers. Pilson, M. E. Q. (1998). *An Introduction to the Chemistry of the Sea.* New Jersey: Prentice Hall.

Table 2.5a Composition of bulk Earth, mantle and core

Element (wt%)	Earth	Mantle	Core
Fe	32.0	6.26	85.5
O	29.7	44	0
Si	16.1	21	6
Mg	15.4	22.8	0
Ni	1.82	0.20	5.2
Ca	1.71	2.53	0
Al	1.59	2.35	0
S	0.64	0.03	1.9
Cr	0.47	0.26	0.9
Na	0.18	0.27	0
P	0.07	0.009	0.20
Mn	0.08	0.10	0.03
C	0.07	0.01	0.20
H	0.03	0.01	0.06
Total	99.86	99.83	99.99

Note: The compositions are based on a model in which the Fe/Ni in the Earth is the same as in chondritic meteorites. Composition of the core is an active area of research – some workers suggest the presence of a small amount of oxygen.

Source: McDonough, W. F. (2004). Compositional model for the Earth's core. In *The Mantle and Core*, ed. R. W. Carlson, vol. 2, *Treatise on Geochemisty*, ed. H. D. Holland and K. K. Turekian. Oxford: Elsevier-Pergamon.

Table 2.5b Composition of the Earth's mantle[a]

Z	Element µg/g except where stated		Z	Element ng/g except where stated	
3	Li	1.6	37	Rb µg/g	0.605
4	Be	0.070	38	Sr µg/g	20.3
5	B	0.26	39	Y µg/g	4.37
6	C	100	40	Zr µg/g	10.81
7	N	2	41	Nb	588
8	O wt%	44.33	42	Mo	39
9	F	25	44	Ru	4.55
11	Na	2590	45	Rh	0.93
12	Mg wt%	22.17	46	Pd	3.27
13	Al	23 800	47	Ag	4
14	Si wt%	21.22	48	Cd	64
15	P	86	49	In	13
16	S	200	50	Sn	138
17	Cl	30	51	Sb	12
19	K	260	52	Te	8
20	Ca	26 100	53	I	7
21	Sc	16.5	55	Cs	18
22	Ti	1280	56	Ba	6750
23	V	86	57	La	686
24	Cr	2520	58	Ce	1786
25	Mn	1050	59	Pr	270
26	Fe wt%	6.3	60	Nd	1327
27	Co	102	62	Sm	431
28	Ni	1860	63	Eu	162
29	Cu	20	64	Gd	571
30	Zn	53.5	65	Tb	105
31	Ga	4.4	66	Dy	711
32	Ge	1.2	67	Ho	159
33	As	0.066	68	Er	465
34	Se	0.079	69	Tm	71.7
35	Br	0.075	70	Yb	462

Table 2.5b *(cont.)*

Z	Element	ng/g	Z	Element	ng/g
71	Lu	71.1	79	Au	0.88
72	Hf	300	80	Hg	6
73	Ta	40	81	Tl	3
74	W	16	82	Pb	185
75	Re	0.32	83	Bi	5
76	Os	3.4	90	Th	83.4
77	Ir	3.2	91	U	21.8
78	Pt	6.6			

Note: *a* The composition is that of the primitive mantle, i.e. the mantle before the onset of crust formation and so represents the silicate component of the Earth. It is a model composition largely based on the premise that the bulk compositon of the Earth is chondritic.

Source: Palme, H. and O'Neill H. St. C. (2004). Cosmochemical estimates of mantle composition. In *The Mantle and Core*, ed. R. W. Carlson, vol. 2, *Treatise on Geochemisty*, ed. H. D. Holland and K. K. Turekian. Oxford: Elsevier-Pergamon.

Table 2.5c Composition of the continental crust

Element	Bulk[R]	Bulk[T]	Upper crust[R]	Middle crust[R]	Lower crust[R]
wt%					
SiO_2	60.6	57.3	66.6	63.5	53.4
TiO_2	0.72	0.9	0.64	0.69	0.82
Al_2O_3	15.9	15.9	15.4	15.0	16.9
FeO_T	6.71	9.1	5.04	6.02	8.57
MnO	0.10	0.18	0.10	0.10	0.10
MgO	4.66	5.3	2.48	3.59	7.24
CaO	6.41	7.4	3.59	5.25	9.59
Na_2O	3.07	3.1	3.27	3.39	2.65
K_2O	1.81	1.1	2.80	2.30	0.61
P_2O_5	0.13		0.15	0.15	0.10
Total	100.11	100.28	100.07	99.99	99.98
µg/g *(asterisk indicates* ng/g *or ppb)*					
Li	17	13	24	12	13
Be	1.9	1.5	2.1	2.3	1.4
B	11	10	17	17	2
N	56		83		34
F	553		557	524	570
S	404		621	249	345
Cl	244		294	182	250
Sc	21.9	30	14	19	31
V	138	230	97	107	196
Cr	135	185	92	76	215
Co	26.6	29	17.3	22	38
Ni	59	105	47	33.5	88
Cu	27	75	28	26	26
Zn	72	80	67	69.5	78
Ga	16	18	17.5	17.5	13
Ge	1.3	1.6	1.4	1.1	1.3
As	2.5	1.0	4.8	3.1	0.2
Se	0.13	0.05	0.09	0.064	0.2

Table 2.5c (*cont.*)

Element	Bulk[R]	Bulk[T]	Upper crust[R]	Middle crust[R]	Lower crust[R]
μg/g (*asterisk indicates* ng/g *or ppb*)					
Br	0.88		1.6		0.3
Rb	49	37	82	65	11
Sr	320	260	320	282	348
Y	19	20	21	20	16
Zr	132	100	193	149	68
Nb	8	11	12	10	5
Mo	0.8	1.0	1.1	0.60	0.6
Ru*	0.6		0.34		0.75
Pd*	1.5	1.0	0.52	0.76	2.8
Ag*	56	80	53	48	65
Cd	0.08	0.10	0.098	0.061	0.10
In	0.05	0.05	0.056		0.05
Sn	1.7	2.5	2.1	1.3	1.7
Sb	0.2	0.2	0.4	0.28	0.10
I	0.7		1.4		0.14
Cs	2	1	4.9	2.2	0.3
Ba	456	250	628	532	259
La	20	16	31	24	8
Ce	43	33	63	53	20
Pr	4.9	3.9	7.1	5.8	2.4
Nd	20	16	27	25	11
Sm	3.9	3.5	4.7	4.6	2.8
Eu	1.1	1.1	1.0	1.4	1.1
Gd	3.7	3.3	4.0	4.0	3.1
Tb	0.6	0.60	0.7	0.7	0.48
Dy	3.6	3.7	3.9	3.8	3.1
Ho	0.77	0.78	0.83	0.82	0.68
Er	2.1	2.2	2.3	2.3	1.9
Tm	0.28	0.32	0.30	0.32	0.24
Yb	1.9	2.2	2.0	2.2	1.5
Lu	0.30	0.30	0.31	0.4	0.25

(*cont.*)

2

Table 2.5c (*cont.*)

Element	Bulk^R	Bulk^T	Upper crust^R	Middle crust^R	Lower crust^R
μg/g *(asterisk indicates* ng/g *or ppb)*					
Hf	3.7	3.0	5.3	4.4	1.9
Ta	0.7	1.0	0.9	0.6	0.6
W	1	1.0	1.9	0.60	0.60
Re*	0.188	0.5	0.198		0.18
Os*	0.041		0.031		0.05
Ir*	0.037	0.10	0.022		0.05
Pt*	1.5		0.5	0.85	2.7
Au*	1.3	3.0	1.5	0.66	1.6
Hg	0.03		0.05	0.008	0.014
Tl	0.50	0.36	0.9	0.27	0.32
Pb	11	8.0	17	15.2	4
Bi	0.18	0.06	0.16	0.17	0.2
Th	5.6	3.5	10.5	6.5	1.2
U	1.3	0.91	2.7	1.3	0.2

Notes: The estimate from Taylor and McLennan is given because of its frequent citation.

Sources: R = Rudnick, R. L. and Gao, S. (2004). Composition of the continental crust. In *The Crust*, ed. R. L. Rudnick, vol. 3, *Treatise on Geochemistry*, ed. H. D. Holland and K. K. Turekian. Oxford: Elsevier-Pergamon. T = Taylor, S. R. and McLennan, S. M. (1985). *The Continental Crust: Its Composition and Evolution*. Oxford: Blackwell, with updates by McLennan (2001) for Rb, Cs and Ta.

Table 2.6a Average compositions (major elements) of common igneous rocks

Oxides (wt%)	Dunite	Harzburgite	Peridotite	Lherzolite	Anorthosite	Gabbro	Basalt
SiO_2	38.29	39.93	42.26	42.52	50.28	50.14	49.20
TiO_2	0.09	0.26	0.63	0.42	0.64	1.12	1.84
Al_2O_3	1.82	2.35	4.23	4.11	25.86	15.48	15.74
Fe_2O_3	3.59	5.48	3.61	4.82	0.96	3.01	3.79
FeO	9.38	6.47	6.58	6.96	2.07	7.62	7.13
MnO	0.71	0.15	0.41	0.17	0.05	0.12	0.20
MgO	37.94	33.18	31.24	28.37	2.12	7.59	6.73
CaO	1.01	2.90	5.05	5.32	12.48	9.58	9.47
Na_2O	0.20	0.31	0.49	0.55	3.15	2.39	2.91
K_2O	0.08	0.14	0.34	0.25	0.65	0.93	1.10
H_2O^+	4.59	4.00	3.91	1.07	1.17	0.75	0.95
H_2O^-	0.25	0.24	0.31	0.03	0.14	0.11	0.43
P_2O_5	0.2	0.13	0.1	0.11	0.09	0.24	0.35
CO_2	0.43	0.09	0.3	0.08	0.14	0.07	0.11
Totals	98.58	95.63	99.46	94.78	99.80	99.15	99.95

(*cont.*)

2

2

wt%	Tholeiite	Alkali basalt	Andesite	Dacite	Tonalite	Diorite and quartz diorite	Granodiorite and quartz monzodiorite
SiO_2	49.58	47.1	57.94	65.01	64.82	54.31	66.37
TiO_2	1.98	2.7	0.87	0.58	0.62	0.95	0.53
Al_2O_3	14.79	15.3	17.02	15.91	16.5	17.76	15.90
Fe_2O_3	3.38	4.3	3.27	2.43	1.30	2.27	1.07
FeO	8.03	8.3	4.04	2.30	3.47	5.31	2.87
MnO	0.18	0.17	0.14	0.09	0.09	0.13	0.08
MgO	7.30	7.0	3.33	1.78	2.28	4.62	1.66
CaO	10.36	9.0	6.79	4.32	4.78	7.8	3.42
Na_2O	2.37	3.4	3.48	3.79	3.50	3.83	3.51
K_2O	0.43	1.2	1.62	2.17	1.74	1.06	2.91
H_2O^+	0.91		0.83	0.91	0.83	1.09	0.68
H_2O^-	0.50		0.34	0.28			
P_2O_5	0.24	0.41	0.21	0.15	0.18	0.25	0.15
CO_2	0.03		0.05	0.06	0.75	0.04	0.07
Totals	100.08	98.88	99.93	99.78	100.86	99.42	99.22

Table 2.6a (*cont.*)

wt%	Rhyolite	Granite	Trachyte	Syenite	Nepheline syenite	Phonolite	Ca-carbonatite
SiO_2	72.82	71.30	61.21	58.58	53.56	56.77	9.09
TiO_2	0.28	0.31	0.70	0.84	0.73	0.59	0.64
Al_2O_3	13.27	14.32	16.96	16.64	20.22	19.76	2.85
Fe_2O_3	1.48	1.21	2.99	3.04	3.45	2.48	4.29
FeO	1.11	1.64	2.29	3.13	2.41	1.63	4.28
MnO	0.06	0.05	0.15	0.13	0.20	0.18	0.71
MgO	0.39	0.71	0.93	1.87	0.76	0.51	6.57
CaO	1.14	1.84	2.34	3.53	2.37	2.32	34.20
Na_2O	3.55	3.68	5.47	5.24	8.43	8.41	1.00
K_2O	4.30	4.07	4.98	4.95	5.68	5.25	1.43
H_2O^+	1.10	0.64	1.15	0.99			
H_2O^-	0.31	0.13	0.47	0.23			
P_2O_5	0.07	0.12	0.21	0.29	0.24	0.16	1.27
CO_2	0.08	0.05	0.09	0.28	0.42	0.61	29.47
Totals	99.96	100.07	99.94	99.74	98.85[a]	99.06[b]	96.55[c]

Notes: a Total includes 0.38% Cl and F. b Includes 0.39% Cl and F. c Includes 0.75 Cl and F. Rock terminology has not necessarily been used previously in exactly the same way as now recommended by the IUGC's Subcommission on the Systematics of Igneous Rocks (2002, see Figs. 2.2–2.6). However, the data above give a good indication of the average compositions of the individual rock types. Ranges of concentrations of oxides can be found in the text by Hyndman. **Sources:** Hyndman, D. W. (1985). *Petrology of Igneous and Metamorphic Rocks*, 2nd edn. New York: McGraw-Hill Book Company. Le Maitre, R. W. (1976). The chemical variability of some common igneous rocks. *Journal of Petrology*, **17**, 589–637.

Table 2.6b Compositions of oceanic basalts (ridge and islands)[a]

Oxide (wt%)	Normal MORB	Enriched MORB	Hawaiian Island	
			Alkali	Tholeiite
SiO_2	50.01	51.28	46.37	51.75
TiO_2	1.11	1.83	2.40	2.07
Al_2O_3	16.31	15.23	14.18	13.81
Fe_2O_3			4.09	2.48
FeO	9.73	9.60	8.91	8.35
MnO	0.14	0.16	0.19	0.17
MgO	8.67	7.43	9.47	7.25
CaO	11.75	10.59	10.33	10.57
Na_2O	2.52	3.08	2.85	2.31
K_2O	0.05	0.53	0.93	0.42
P_2O_5	0.08	0.26	0.28	0.24
Total	100.37	99.99	100.00	99.42

Element Z ($\mu g/g$)				
21 Sc	44	36	27.5	29.5
23 V	281	288		
24 Cr	251		447	350
28 Ni	119	91		
29 Cu	68			
31 Ga			19	20
37 Rb	0.38	8.85	22	6.9
38 Sr	94	181	500	344
39 Y	25	29	22	23
40 Zr	57	134	148	129
41 Nb	1.07	11.2	25	9
55 Cs	0.006	0.08		
56 Ba	6.11	123	300	
57 La	1.88	11.5	18.8	9.38
58 Ce	5.99	26	43.0	25.0

Table 2.6b (*cont.*)

Element Z (μg/g)	Normal MORB	Enriched MORB	Hawaiian Island	
			Alkali	Tholeiite
60 Nd	6.07	17.1		
62 Sm	2.22	4.38	5.35	4.80
63 Eu	0.9	1.54	1.76	1.70
64 Gd	3.5	5.26		
65 Tb			0.88	0.81
66 Dy	4.46	5.24		
68 Er	2.57	2.83		
70 Yb	2.72	2.68	1.88	2.00
71 Lu			0.28	0.29
72 Hf	2.9	2.06	4.0	3.48
73 Ta	0.13	0.77		
82 Pb	0.19	0.95	7	<1
90 Th	0.09	1.07	1.94	0.56
92 U	0.03	0.28		

Notes: *a* MORB = mid-ocean ridge basalt. Here, the normal and enriched MORB are composite analyses of basalts from the Mid Atlantic Ridge. It should be noted that compositions of different MORBs can show significant variation; the upper mantle is not compositionally homogeneous. Basalts are divided into alkali basalt (which has CIPW normative nepheline) and subalkali basalt (which does not). MORB and tholeittic basalt are subalkali basalts.

Sources: Klein, E. M. (2004). Geochemistry of the igneous oceanic crust. In *Treatise on Geochemistry*, vol. 3, ed. H. D. Holland and K. K. Turekian. Oxford: Elsevier-Pergamon, pp. 433–463. Basaltic Volcanism Study Project. (1981). *Basaltic Volcanism on the Terrestrial Planets*. New York: Pergamon Press.

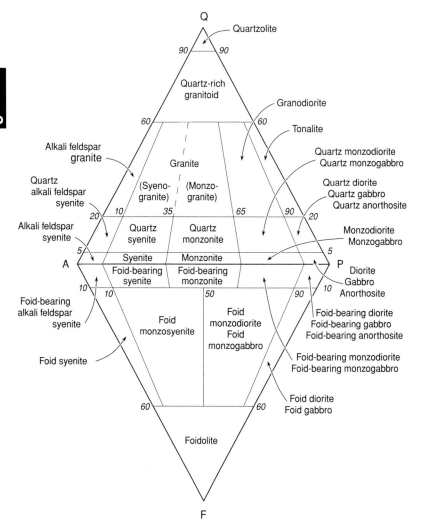

Figure 2.2 Classification diagram for plutonic igneous rocks based on the modal proportions of quartz Q, alkali feldspar A, plagioclase P and feldspathoid (= foid) F. It is the scheme for these rocks as recommended by the International Union of Geological Sciences, Subcommission on the Systematics of Igneous Rocks. It is applicable to rocks in which the mafic mineral content is less than 90%.

Source: Le Maitre, R. W. *et al.* (2002). *Igneous Rocks. A Classification and Glossary of Terms.* 2nd edn. Cambridge: Cambridge University Press, with permission.

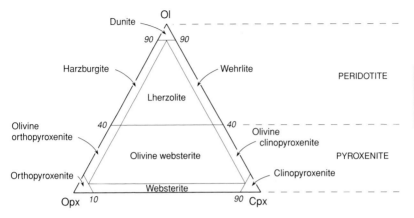

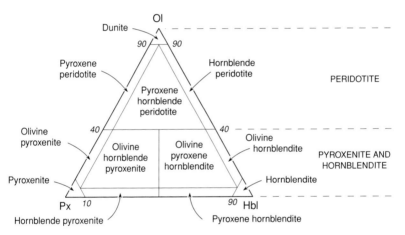

Figure 2.3 Classification diagrams for ultramafic igneous rocks based on the modal proportions of olivine Ol, orthopyroxene Opx, clinopyroxene Cpx, pyroxene Px and hornblende Hbl. It is the scheme for these rocks as recommended by the International Union of Geological Sciences, Subcommission on the Systematics of Igneous Rocks.

Source: as for Fig. 2.2 with permission.

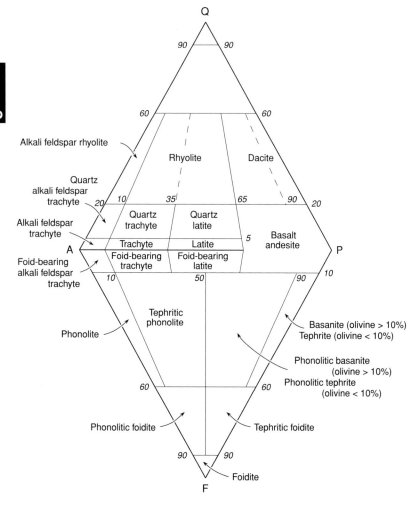

Figure 2.4 Classification diagram for volcanic rocks based on the modal proportions of quartz, Q, alkali feldspar A, plagioclase P and feldspathoid (=foid) F. It is the scheme for these rocks as recommended by the International Union of Geological Sciences, Subcommission on the Systematics of Igneous Rocks.

Source: as for Fig. 2.2 with permission.

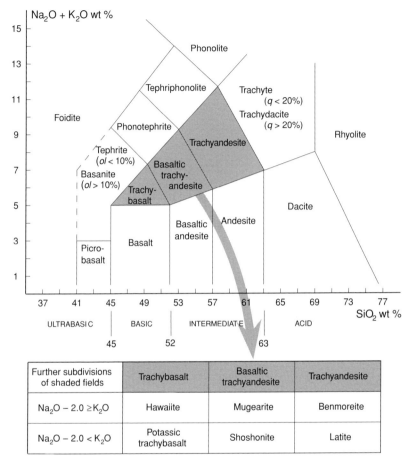

Figure 2.5 Chemical classification of volcanic rocks using the total alkali and silica contents. The line between the foidite field and the basanite–tephrite field is dashed to indicate that further criteria must be used to separate these types.

Abbreviations: ol = normative olivine; q = normative percentage of quartz Q as $(100\{Q/(Q + or + ab + an)\}$. It is the scheme for these rocks as recommended by the International Union of Geological Sciences, Subcommission on the Systematics of Igneous Rocks.

Source: as for Fig. 2.2 with permission.

Table 2.7 Grain-size scale for sediments and sedimentary rocks

Length (mm)	ϕ^a	Descriptor	Class	Sediment or rock name
>4096	<−12		Block	Mega-conglomerate
2048–4096	−11 to −12	Very coarse		
1024–2048	−10 to −11	Coarse		
512–1024	−9 to −10	Medium	Boulder	
256–512	−8 to −9	Fine		
128–256	−7 to −8	Coarse	Cobble	Gravel conglomerate
64–128	−6 to −7	Fine		
32–64	−5 to −6	Very coarse		
16–32	−4 to −5	Coarse		
8–16	−3 to −4	Medium	Pebble	
4–8	−2 to −3	Fine		
2–4	−1 to −2		Granule	
1–2	0 to −1	Very coarse		
0.50–1	1 to 0	Coarse		
0.25–0.5	2 to 1	Medium	Sand	Sand sandstone
0.125–0.25	3 to 2	Fine		
0.063–0.125	4 to 3	Very fine		
0.031–0.063	5 to 4	Coarse		
0.015–0.031	6 to 5	Medium		
0.008–0.015	7 to 6	Fine	Silt	Silt siltstone
0.004–0.008	8 to 7	Very fine		
<0.004	>8		Clay	Clay claystone

Note: a an arithmetic scale of phi (ϕ) units has been introduced to aid certain calculations, where $\phi = -\log_2 d$, where d is length (mm).

Source: Tucker, M. E. (2001). *Sedimentary Petrology. An Introduction to the Origin of Sedimentary Rocks.* Oxford: Blackwell Science, with permission.

Table 2.8 Mean composition of sandstone types and of average sandstone

Oxide	Quartz arenite	Lithic arenite	Graywacke	Arkose	Average sandstone[a]
SiO_2	95.4	66.1	66.7	77.1	78.66
TiO_2	0.2	0.3	0.6	0.3	0.25
Al_2O_3	1.1	8.1	13.5	8.7	4.78
Fe_2O_3	0.4	3.8	1.6	1.5	1.08
FeO	0.2	1.4	3.5	0.7	0.3
MnO		0.1	0.1	0.2	Trace
MgO	0.1	2.4	2.1	0.5	1.17
CaO	1.6	6.2	2.5	2.7	5.52
Na_2O	0.1	0.9	2.9	1.5	0.45
K_2O	0.2	1.3	2	2.8	1.32
P_2O_5		0.1	0.2	0.1	0.08
H_2O^+	0.3	3.6	2.4	0.9	1.33
H_2O^-		0.7	0.6		0.31
CO_2	1.1	5.0	1.2	3	5.04
SO_3			0.3		0.07
Total	100.7	100.0	100.2	100.0	100.36

Note: a Composite analysis of 253 sandstones.

Source: Pettijohn, F. J., Potter, P. E. and Siever, R. (1987). *Sand and Sandstone.* 2nd edn. New York: Springer-Verlag.

Table 2.9 Coal classification according to rank

Rank	Refl.[a] (Rm$_{oil}$)	Organic[b] C (%)	Atomic H/C	Atomic O/C	Characteristics
Peat	0.2	50–60	0.80–1.40	0.4–0.7	Identifiable plant fragments; abundant cellulose
Lignite	0.3	60–70	0.80–1.20	0.2–0.5	Some identifiable plant fragments; only a residue of cellulose; dehydration
Sub-bituminous					
C	0.4				Gelification; no cellulose; loss of methoxyl functionality in lignin
B	0.4–0.5	70–80	0.75–1.00	0.15	
A	0.5				
High volatile bituminous					
C	0.6				Condensation of aromatic structures; loss of H and O; peak of oil-generation zone.
B	0.7	80–90	0.60–0.75	0.10	
A	0.8–1.1				
Medium volatile bituminous	1.1–1.5	80–90	0.50–0.75	0.05–0.10	
Low volatile bituminous	1.5–1.9	85–90	0.50–0.75	0.05–0.10	

Table 2.9 (*cont.*)

Rank	Refl.[a] (Rm$_{oil}$)	Organic[b] C (%)	Atomic H/C	Atomic O/C	Characteristics
Semi-anthracite	2.0	90	0.5	0.05	
Anthracite	3.0	>90	0.25–0.50	<0.05	Condensation of aromatic structures; loss of H and O nearly completely; graphitisation
Meta-anthracite	>4.0	>90	0.25	<0.05	

Notes: a Vitrinite reflectance measured in oil.
b Organic carbon, weight per cent, in dry, ash-free material.
There are numerous classificatory systems for coal; the one presented here gives chemical and other data of interest to earth scientists. It is reproduced from Orem, W. H. and Finkelman, R. B. (2004). Coal formation and geochemistry. In *Sediments, Diagenesis, and Sedimentary Rocks,* ed. F. T. Mackenzie, vol. 7, *Treatise on Geochemistry,* ed. H. D. Holland and K. K. Turekian. Oxford: Elsevier-Pergamon, with permission.

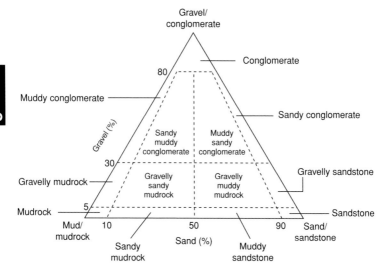

Figure 2.6 Classification diagram for sand–gravel–mud mixtures, with the terms for sediment and for rock.

Source: Tucker, M. E. (2001). *Sedimentary Petrology. An Introduction to the Origin of Sedimentary Rocks*, 3rd edn. Oxford: Blackwell Science Ltd., with permission of the publisher.

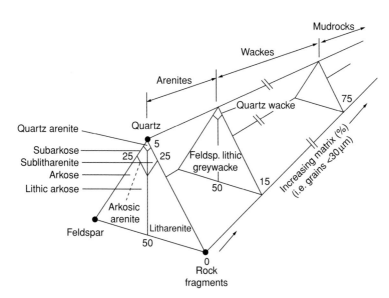

Figure 2.7 Classification of sandstones.

Source: see Fig. 2.6, with permission.

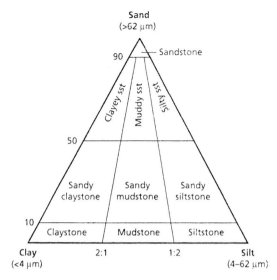

Figure 2.8 Classification diagram for siliciclastic sediments based on sand, silt and clay content.

Source: see Fig. 2.6, with permission.

Original components not bound together during deposition				Original components bound together	Depositional texture not recognizable	Original components not organically bound during deposition		Original components organically bound during deposition		
Contains lime mud			Lacks mud and is grain supported			>10% grains >2 mm		Organisms act as baffles	Organisms encrust and bind	Organisms build a rigid framework
Mud-supported		Grain supported				Matrix supported	Supported by >2 mm components			
Less than 10% grains	More than 10% grains				Crystalline carbonate					
Mudstone	Wackestone	Packstone	Grainstone	Boundstone	Crystalline	Floatstone	Rudstone	Bafflestone	Bindstone	Framestone

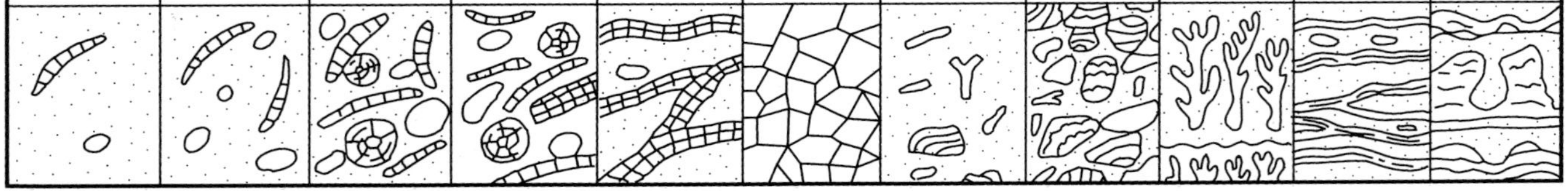

Figure 2.9 Classification of limestones based on depositional texture. The scheme is modified after one by R. J. Dunham.

Source: see Fig. 2.6, with permission.

Table 2.10 Types of metamorphism

Name	Characteristic or association
Regional	Over a wide area; large volume
Local	Limited area; limited volume
Orogenic	Regional and of orogenic belt
Burial	Regional. Deeply buried
Ocean floor	Regional or local and related to steep geothermal gradient as near spreading centres
Dislocation	Local. Associated with fault or shear zone
Impact	Local. Shock wave from impact (e.g. meteorite)
Contact	Local. Around magma body
Pyrometamorphic	A type of contact metamorphism involving very high temperature and low pressure
Hydrothermal.	Local. Caused by hot water-rich fluids
Hot slab	Local. Beneath emplaced hot tectonic body
Combustion	Local. Spontaneous combustion of coals etc.
Lightning	Local. Lightning strike

Source. Based on: Fettes, D. and Desmons, J. (2007). *Metamorphic Rocks. A Classification and Glossary of Terms.* Cambridge: Cambridge University Press.

Table 2.11 Metamorphic facies and mineral paragenesis

	Facies	Mineral paragenesis
1	Zeolite	Laumontite, heulandite etc.
2	Subgreenschist	Prehnite–pumpellyite, pumpellyite–actinolite, prehnite–actinolite
3	Greenschist	Actinolite–albite–epidote-chlorite
4	Epidote–amphibolite	Hornblende–albite–epidote
5	Amphibolite	Hornblende–plagioclase
6	Pyroxene–hornfels	Clinopyroxene–orthopyroxene–plagioclase (olivine stable with plagioclase)
7	Sanidinite	High-T varieties and polymorphs of minerals
8	Glaucophane-schist	Glaucophane–epidote-(garnet)
	or blue-schist	Glaucophane–lawsonite, glaucophane–lawsonite–jadeite
9	Eclogite	Omphacite–garnet–quartz
10	Granulite	Clinopyroxene–orthopyroxene–plagioclase (olivine not stable with plagioclase)

Note: The International Union of Geological Sciences, Subcommission on the Systematics of Metamorphic Rocks, recommends that the above 10 facies be adopted for general use. See also Fig. 2.10: Pressure–temperature diagram of the main metamorphic facies.

Source: Fettes, D. and Desmons, J., eds. (2007). *Metamorphic Rocks. A Classification and Glossary of Terms*. Cambridge: Cambridge University Press. This book gives recommendations on how to name a metamorphic rock, including use of specific terms.

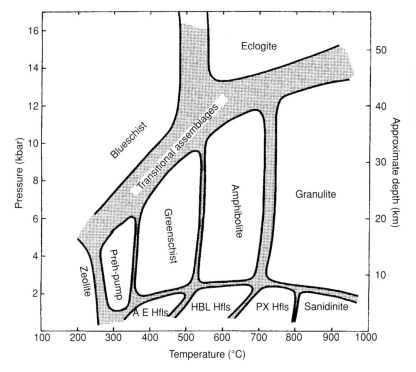

Figure 2.10 Pressure–temperature diagram showing the fields of the main metamorphic facies. Hfls = hornfels, AE = albite–epidote, HBL = hornblende, PX = pyroxene and Preh–pump = prehnite–pumpellyite.

Source: Yardley, B. W. (1989). *An Introduction to Metamorphic Petrology.* Harlow: Longman, with permission of the author.

3 Geophysics

The data given in this chapter are for conditions of standard temperature and pressure unless specified otherwise.

Data of relevance to geophysics can be found in other chapters:

- Chapter 1 includes data on the planets, moon and meteorites,
- Chapter 2 includes data on the properties of the Earth as well as information on surface features,
- Information on the age of the Earth and geological time scales can be found in Chapter 7.
- Chapter 7 also includes data on key events during the Earth's history, and data on recent volcanic events and earthquakes.
- Chapter 11 includes data on volcanism, earthquakes and tsunamis.

Table 3.1 Physical properties of rocks

Rock	Density (g cm^{-3}) Range	Average	Young's modulus (Gpa)	Poisson's ratio	Porosity% Range	Average
Sedimentary rocks						
Chalk	1.4–2.6	1.9			10–75	50
Limestone	2.0–2.7	2.4	50–80	0.15–0.3	3–41	15
Dolomite	2.3–2.8	2.6	50–90	0.1–0.4	0–32	13
Sandstone	2.1–2.7	2.4	10–60	0.1–0.3	2–32	17
Shale	1.8–3.2	2.4	10–70	0.1–0.2		
Igneous rocks						
Rhyolite	2.4–2.7	2.5				
Andesite	2.4–2.8	2.6				
Granite	2.5–2.8	2.6	40–70	0.20–0.25		
Diorite	2.7–3.0	2.9	60–80	0.25–0.30		
Basalt	2.7–3.3	3.0	60–80	0.20–0.25		
Gabbro	2.7–3.5	3.0	60–100	0.15–0.20		
Peridotite	2.8–3.4	3.2				
Metamorphic rocks						
Quartzite	2.5–2.7	2.6				
Schist	2.4–2.9	2.6				
Marble	2.6–2.9	2.8	30–80	0.2–0.3		
Slate	2.7–2.9	2.8				
Serpentine	2.4–3.1	2.8				
Gneiss	2.6–3.0	2.8	40–60	0.15–0.25		
Amphibolite	2.9–3.0	3.0		0.4		
Eclogite	3.2–3.5	3.4				

Note: Densities of minerals can be found in Chapter 9.

Sources: Mavko, G., Mukerji, T. and Dvorkin, J. (1998). *The Rock Physics Handbook.* Cambridge: Cambridge University Press. Telford, W. M., Geldart, L. P. and Sheriff, R. E. (1990). *Applied Geophysics*, 2nd edn. Cambridge: Cambridge University Press. Turcotte, D. L. and Schubert, G. (2002). *Geodynamics*, 2nd edn. Cambridge: Cambridge University Press.

Table 3.2 Moduli and wave velocities of selected minerals

Mineral	Density ρ (g cm^{-3})	Bulk modulus K_S (GPa)	Shear modulus μ (GPa)	Velocities[a]		Poisson's ratio σ
				V_P (km s^{-1})	V_S (km s^{-1})	
Non-silicates						
Diamond	3.51	584.8	346.3	17.27	9.93	0.25
Iron	7.87	172.7	83.1	6.00	3.25	0.29
Nickel	8.91	185.6	88.2	5.83	3.15	0.29
Halite	2.16	24.7	14.4	4.51	2.58	0.26
Periclase	3.58	162.8	129.4	9.68	6.01	0.19
Corundum	3.99	251.9	162	10.83	6.37	0.24
Rutile	4.26	217.1	108.1	9.21	5.04	0.29
Spinel	3.63	203.1	116.1	9.93	5.65	0.26
Perovskite	4.04	177.0	104.0	8.84	5.07	0.25
Pyrite	4.93	147.4	132.5	8.10	5.18	0.15
Sphalerite	4.08	75.2	32.3	5.36	2.81	0.31
Barite	4.51	54.5	23.8	4.37	2.3	0.31
Celestine	3.96	81.9	21.4	5.28	2.33	0.38
Anhydrite	2.98	56.1	29.1	5.64	3.13	0.28
Calcite	2.71	70.8	30.3	6.41	3.35	0.31
Siderite	3.96	123.7	51.0	6.96	3.59	0.32
Dolomite	2.87	80.2	48.8	7.11	4.12	0.25
Aragonite	2.92	44.8	38.8	5.75	3.64	0.16
Apatite	3.22	83.9	60.7	7.15	4.34	0.21
Silicates						
Quartz	2.65	37.0	44.7	6.05	4.11	0.07
Albite	2.63	75.6	25.6	6.46	3.12	0.35
Forsterite	3.21	128.2	80.5	8.57	5.00	0.24
Almandine garnet	4.18	176.3	95.2	8.51	4.77	0.27
Pyrope garnet	3.56	176.6	89.6	9.12	5.02	0.28
Zircon	4.56	19.8	19.7	3.18	2.08	0.13
Epidote	3.40	106.5	61.1	7.43	4.24	0.26
Enstatite	3.20	107.5	75.4	8.06	4.85	0.22

Table 3.2 (*cont.*)

Mineral	Density ρ (g cm^{-3})	Bulk modulus K_S (GPa)	Shear modulus μ (GPa)	Velocities[a] V_P (km s^{-1})	V_S (km s^{-1})	Poisson's ratio σ
Diopside	3.31	111.2	63.7	7.70	4.39	0.26
Muscovite	2.79	52.1	31.4	5.78	3.33	0.25
Phlogopite	2.80	58.5	40.1	6.33	3.79	0.22
Biotite	3.05	59.7	42.3	6.17	3.73	0.21
Kaolinite	1.58	1.5	1.4	1.44	0.93	0.14
'Perovskite' MgSiO$_3$	4.11	246.4	184.2	10.94	6.69	0.20
Akimokoite MgSiO$_3$	3.80	212.0	132.0	10.10	5.89	0.24

Note: *a* V_P and V_S are velocities of the longitudinal (P) and transverse (S) seismic waves respectively.

Sources: Mavko, G., Mukerji, T. and Dvorkin, J. (1998). *The Rock Physics Handbook.* Cambridge: Cambridge University Press. Poirier, J. P. (2000). *Introduction to the Physics of the Earth's Interior,* 2nd edn. Cambridge: Cambridge University Press.

Table 3.3 Average magnetic susceptibilities of selected rocks and minerals[a]

Material	Magnetic susceptibility	Material	Magnetic susceptibility
Sedimentary rocks		*Minerals*	
Limestone	0.3	Quartz	−0.01
Sandstone	0.4	Gypsum	−0.01
Shale	0.6	Clays	0.2
Metamorphic rocks		Chalcopyrite	0.4
Schist	1.4	Sphalerite	0.7
Quartzite	4	Cassiterite	0.9
Slate	6	Pyrite	1.5
Igneous rocks		Arsenopyrite	3
Granite	2.5	Hematite	6.5
Gabbro	70	Chromite	7
Andesite	160	Pyrrhotite	1500
Basalt	70	Ilmenite	1800
Diorite	85	Magnetite	6000
Peridotite	150		

Note: *a* Magnetic susceptibility is a constant of proportionality in the relationship between induced magnetisation and the strength of an applied field. It is, therefore, dimensionless.

Source: Telford W. M., Geldart, L. P. and Sheriff, R. E. (1990). *Applied Geophysics*, 2nd edn. Cambridge: Cambridge University Press.

Table 3.4a Resistivities of waters

Water	Resistivity range (Ωm)
Meteoric waters	$30–10^3$
Surface waters	$0.1 − 3 \times 10^3$
Soil waters	100 (av.)
Natural waters	0.5–150
Sea water	0.2–0.3

Notes: (av.) = average value. Resistivity is proportional to the resistance and area of the material and inversely proportional to its length. The reciprocal of resistivity is the conductivity of the material. Units here are ohm-metre.

Sources: Jones, E. J. W. (1999). *Marine Geophysics*. Chichester: John Wiley. Telford, W. M., Geldart, L. P. and Sheriff, R. E. (1990). *Applied Geophysics*, 2nd edn. Cambridge: Cambridge University Press.

Table 3.4b Resistivities of minerals

Mineral	Resistivity range (Ωm)
Non-silicates	
Native copper	$1.2 \times 10^{-8} - 3 \times 10^{-7}$
Chromite	$1 - 1 \times 10^6$
Bauxite	$200 - 6 \times 10^3$
Cassiterite	$4 \times 10^{-4} - 1 \times 10^4$
Hematite	$3.5 \times 10^{-3} - 10^7$
Ilmenite	$1 \times 10^{-3} - 50.0$
Magnetite	$1.5 \times 10^{-5} - 7.5 \times 10^5$
Rutile	$30 - 1000$
Uraninite	$1 - 200$
Calcite	2×10^{12} (av.)
Siderite	70 (av.)
Anhydrite	1×10^9 (av.)
Bornite	$2.5 \times 10^{-5} - 0.5$
Chalcocite	$3 \times 10^{-5} - 0.6$
Chalcopyrite	$1.2 \times 10^{-5} - 0.3$
Cinnabar	2×10^7 (av.)
Galena	$1 \times 10^{-5} - 3 \times 10^2$
Molybdenite	$1 \times 10^{-3} - 1 \times 10^6$
Pyrite	$1.0 \times 10^{-5} - 1.5$
Pyrrhotite	$2 \times 10^{-6} - 5 \times 10^{-2}$
Arsenopyrite	$2 \times 10^{-5} - 15$
Sphalerite	$1.5 - 1 \times 10^7$
Fluorite	8×10^{13} (av.)
Silicates and Quartz	
Quartz	$4 \times 10^{10} - 2 \times 10^{14}$
Hornblende	$200 - 1 \times 10^6$
Muscovite	$9 \times 10^2 - 1 \times 10^{14}$
Biotite	$200 - 1 \times 10^6$

Notes and sources: see Table 3.4a.

Table 3.4c Resistivities of rocks and sediments

Material	Resistivity range (Ωm)
Clays	$1-100$
Shales	$0.4-2 \times 10^3$
Limestone	$1.3 \times 10^3 - 8.4 \times 10^6$
Sandstone	$1-6.4 \times 10^8$
Quartzite	$10-2 \times 10^8$
Slate	$2 \times 10^{-4} - 2 \times 10^7$
Marble	$100-2.5 \times 10^8$
Schist	$20-10^4$
Gneiss	$2 \times 10^3 - 3 \times 10^6$
Hornfels	$8 \times 10^3 - 6 \times 10^7$
Basalt	$10-1.3 \times 10^7$
Andesite	$100-4.5 \times 10^4$
Syenite	$100-10^6$
Gabbro	$100-10^6$
Peridotite	$3 \times 10^3 - 6.5 \times 10^3$
Granite	$160-3.6 \times 10^6$

Notes and sources: see Table 3.4a.

Table 3.5 Average thermal conductivities of minerals[a]

Mineral	Conductivity (W m^{-1} K^{-1})	Mineral	Conductivity (W m^{-1} K^{-1})
Elements		Feldspar–anorthite	2.4
Graphite	200	Quartz	8.4
Diamond	545	*Non-Silicates*	
Silicates		Magnetite	6.5
Olivine–forsterite	4.8	Hematite	12.8
Olivine–fayalite	3.5	Ilmenite	1.9
Garnet–almandine	3.5	Chromite	2.4
Garnet–grossularite	5.4	Spinel	11.8
Zircon	4.7	Rutile	8.9
Titanite	2.3	Corundum	26.5
Kyanite	11.2	Pyrite	26.8
Andalusite	7.1	Pyrrhotite	4.1
Sillimanite	9.9	Galena	2.5
Epidote	2.8	Baryte	2
Pyroxene–enstatite	4.5	Anhydrite	5.1
Pyroxene–augite	4.4	Gypsum	2.7
Pyroxene–jadeite	5.6	Calcite	3.6
Hornblende	2.8	Aragonite	2.3
Mica–muscovite[a]	2.3	Magnesite	7.3
Mica–biotite[a]	1.9	Siderite	3
Mica–talc[a]	4.9	Dolomite	5.1
Mica–chlorite[a]	4.5	Apatite	1.3
Mica–serpentine[a]	2.7	Halite	6
Feldspar–orthoclase	2.4	Sylvite	6.7
Feldspar–microcline	2.3	Fluorite	9.4
Feldspar–albite	2.2		

Note: *a* Thermal conductivity values are dependent on several factors such as temperature and pressure and also on the nature of the mineral including such factors as degree of anisotropy, imperfections and whether or not the measurements were made on a single crystal or an aggregate. Micas, for example, show significant variation in conductivity with direction of the heat flow. The data in this table are compiled from the source below by taking the average of the available measurements. Most are for temperatures within the range 20–35 °C. The reader is referred to the source for further details.

Source: Clauser, C. and Huenges, E. (1995). Thermal conductivity of rocks and minerals. In *Rock Physics and Phase Relations. A Handbook of Physical Constants,* ed. T. J. Ahrens. Washington, DC: American Geophysical Union, pp. 105–126.

Table 3.6 Average thermal conductivities of rocks[a]

Rock	Conductivity (W m^{-1} K^{-1})	Rock	Conductivity (W m^{-1} K^{-1})
Sedimentary Rocks		*Igneous Rocks*	
Mudstone	2.0	Basalt	1.7
Shale	2.1	Gabbro	2.6
Siltstone	2.7	Dunite	4.1
Sandstone	3.7	Andesite	2.3
Limestone	3.4	Diorite	3.0
Dolomite	4.7	Granodiorite	2.6
Metamorphic Rocks		Rhyolite	2.6
Amphibolite	3.3	Granite	3.5
Serpentinite	3.5		
Quartzite	5.0		
Marble	2.8		
Gneiss	3.1		

Note: *a* Thermal conductivities of a particular rock type can vary by a factor of up to 2 (or very occasionally more) depending on the physical nature of the rock. The averages above are therefore to be taken as guideline figures only.

Sources: Jessop, A. M. (1990). *Thermal Geophysics*. Amsterdam: Elsevier. Turcotte, D. L. and Schubert, G. (2002). *Geodynamics*, 2nd edn. Cambridge: Cambridge University Press.

Table 3.7 Thermal expansion coefficients of minerals

Mineral	Volume coefficient[a] $\times 10^{-6}$	Temperature range (K)	Mineral	Volume coefficient[a] $\times 10^{-6}$	Temperature range (K)
Oxides			Garnets:		
Chromite	9.9	293–1273	Almandine	15.8	294–1044
Magnetite	20.6	293–843	Andradite	20.6	294–963
Spinel	24.9	293–873	Grossularite	16.4	292–980
Periclase	31.6	303–1273	Pyrope	19.9	283–1031
Rutile	28.9	298–1883	Spessartine	17.2	292–973
Carbonates			Kyanite	25.1	298–1073
Calcite	3.8	297–1173	Perovskite ($MgSiO_3$)	22.0	298–381
Magnesite	18.2	297–773	Pyroxenes:		
Siderite	26.9	297–1073	Diopside	33.3	297–1273
Sulphides			Enstatite	24.1	293–1073
Pyrite	25.7	293–673	Hedenbergite	29.8	297–1273
Galena	58.1	293–873	Jadeite	24.7	297–1073
Barite	63.7	298–1158	Spinel, γ-Mg_2SiO_4	18.9	297–1023
Silicates			Sillimanite	13.3	298–1273
Andalusite	22.8	298–1273	Silica:		
Cordierite	2.6	298–873	Quartz	24.3	298–773
Fayalite	26.1	298–1123	Coesite	6.9	293–1273
Forsterite	28.2	298–1273	Stishovite	16.4	291–873
Feldspars:			Zircon	12.3	292–1293
Orthoclase	9.7	293–1273			
Albite	15.4	293–1273			
Anorthite	14.1	293–1273			

Note: *a* The coefficient is for volume expansion when it is independent of temperature and is of the form $V_T = V_{T_r} \exp[\alpha(T - T_r)]$ where V_{T_r} is the volume at the reference temperature T_r, and α is the expansion coefficient.

Source: Fei, Y. (1995). Thermal expansion. In *Mineral Physics and Crystallography. A Handbook of Physical Constants*, ed. T. H. Ahrens. Washington, DC: American Geophysical Union.

Table 3.8 Diffusion parameters in selected minerals[a]

Mineral	Diffusing species	Orientation[b]	Temperature range (K)	D_0 $(m^2\ s^{-1})$	Q $(kJ\ mol^{-1})$
Forsterite	Fe	c-axis	1173–1473	4.1×10^{-7}	246
	Fe	c-axis	1398–1573	3.1×10^{-6}	257
	Fe	c-axis	1273–1398	8×10^{-11}	133
	Fe	a-axis	1398–1473	1.3×10^{-5}	283
	Fe		1273–1473	4.2×10^{-10}	162
	Mg		1573–1773	4.1×10^{-7}	277
	Ni	c-axis	1422–1507	1.1×10^{-9}	193
	O		1548–1898	3.5×10^{-7}	372
	O	c-axis	1548–1898	1.3×10^{-6}	398
	Fe–Mg[c]	c-axis	1473–1673	2×10^{-6}	274
	Fe–Mg[c]	b-axis	1473–1673	4.4×10^{-5}	331
Fayalite	Mg	c-axis	1398–1473	2.9×10^{-6}	246
Diopside	O	c-axis	973–1523	1.5×10^{-10}	226
Hornblende	He		453–1123	4.6×10^{-6}	131
Orthoclase	Na		773–1073	8.9×10^{-4}	221
	K		772–1073	16.1×10^{-4}	285
	Ar		673–1173	6.6×10^{-7}	122
Sanidine	He		453–1123	4.6×10^{-8}	94
Albite	O		713–1078	4.5×10^{-9}	155
	Na		573–1073	5.7×10^{-7}	176
	K		773–1073	1.1×10^{-9}	159
	Ar		673–1123	1.6×10^{-7}	138
Almandine	O		1073–1273	6.0×10^{-9}	301
Pyrope	Mg		1023–1173	9.8×10^{-9}	239
Quartz	O		1143–1453	1.1×10^{-14}	195

Notes: *a* The data relate to the Arrhenius equation for diffusion: $D = D_0 \exp(-Q/RT)$ where D is the diffusion rate, D_0 the frequency factor, which can often be treated as a constant, Q the activation energy, T the temperature (K) and R the gas constant.
b Orientation is given where available, otherwise it is assumed to be bulk diffusion.
c Interdiffusion.

Sources: Brady, J. B. (1995). Diffusion data for silicate minerals. In *Mineral Physics and Crystallography. A Handbook of Physical Constants*, ed. T. J. Ahrens. Washington, DC: American Geophysical Union, pp. 269–290. Freer, R. (1981). Diffusion in silicate minerals and glasses: a data digest and guide to the literature. *Contributions to Mineralogy and Petrology*, **76**, 440–454. Henderson, P. (1982). *Inorganic Geochemistry*. Oxford: Pergamon Press. Lasaga, A. C. (1981). The atomistic basis of kinetics: defects in minerals. In *Kinetics of Geochemical Processes*, ed. A. C. Lasaga and R. J. Kirkpatrick, *Reviews in Mineralogy*, vol. 8. Mineralogical Society of America, pp. 261–319. Ozima, M. and Podosek, F. A. (2002). *Noble Gas Geochemistry*, 2nd edn. Cambridge: Cambridge University Press.

Table 3.9 The Preliminary Reference Earth Model (PREM)[a]

Radius (km)	Depth (km)	Pressure P (Gpa)	Density ρ (g cm^{-3})	Velocity V_P (km s^{-1})	Velocity V_S (km s^{-1})	Bulk modulus K_S (Gpa)	Shear modulus μ (Gpa)	Poisson's ratio σ	Gravity acceleration g (cm s^{-2})
0	6371	363.9	13.09	11.26	3.67	1425.3	176.1	0.44	0
100	6271	363.6	13.09	11.26	3.67	1424.8	175.9	0.44	37
200	6171	362.9	13.08	11.26	3.66	1423.1	175.5	0.44	73
300	6071	361.7	13.07	11.25	3.66	1420.3	174.9	0.44	110
400	5971	360.0	13.05	11.24	3.65	1416.4	173.9	0.44	146
500	5871	357.9	13.03	11.22	3.64	1411.4	172.7	0.44	182
600	5771	355.3	13.01	11.21	3.63	1405.3	171.3	0.44	217
700	5671	352.2	12.98	11.18	3.61	1398.1	169.6	0.44	255
800	5571	348.7	12.95	11.16	3.60	1389.8	167.6	0.44	291
900	5471	344.7	12.91	11.14	3.58	1380.5	165.4	0.44	326
1000	5371	340.2	12.87	11.11	3.56	1370.1	163.0	0.44	362
1100	5271	335.4	12.83	11.07	3.54	1358.6	160.3	0.44	397
1200	5171	330.0	12.77	11.03	3.51	1346.2	157.4	0.44	432
1221	5150	328.9	12.76	11.02	3.50	1343.4	156.7	0.44	440
1221	5150	328.9	12.17	10.36	0	1304.7	0	0.5	440
1300	5071	324.5	12.13	10.31	0	1288.8	0	0.5	464

1400	4971	318.7	12.07	10.25	0	1267.9	0	0.5	494
1500	4871	312.6	12.01	10.19	0	1246.4	0	0.5	524
1600	4771	306.1	11.95	10.12	0	1224.2	0	0.5	555
1700	4671	299.3	11.88	10.05	0	1201.3	0	0.5	586
1800	4571	292.2	11.81	9.99	0	1177.5	0	0.5	617
1900	4471	284.8	11.73	9.91	0	1152.9	0	0.5	647
2000	4371	277.0	11.65	9.83	0	1127.3	0	0.5	677
2100	4271	269.0	11.57	9.75	0	1100.9	0	0.5	707
2200	4171	260.7	11.48	9.67	0	1073.5	0	0.5	736
2300	4071	252.1	11.39	9.58	0	1045.1	0	0.5	766
2400	3971	234.2	11.29	9.48	0	1015.8	0	0.5	794
2500	3871	234.2	11.19	9.38	0	988.5	0	0.5	822
2600	3771	224.8	11.08	9.28	0	954.2	0	0.5	850
2700	3671	215.3	10.97	9.17	0	922.0	0	0.5	877
2800	3571	205.6	10.85	9.05	0	888.9	0	0.5	904
2900	3471	195.7	10.73	8.93	0	855.0	0	0.5	930
3000	3371	185.6	10.60	8.80	0	820.2	0	0.5	956
3100	3271	175.4	10.47	8.66	0	784.6	0	0.5	981
3200	3171	165.1	10.33	8.51	0	748.4	0	0.5	1005
3300	3071	154.7	10.18	8.36	0	711.6	0	0.5	1028
3400	2971	144.2	10.02	8.19	0	674.3	0	0.5	1051

(cont.)

3

Table 3.9 (*cont.*)

Radius (km)	Depth (km)	Pressure P (GPa)	Density ρ (g cm^{-3})	Velocity V_P (km s^{-1})	Velocity V_S (km s^{-1})	Bulk modulus K_S (GPa)	Shear modulus μ (GPa)	Poisson's ratio σ	Gravity acceleration g (cm s^{-2})
3480	2891	135.8	9.90	8.06	0	644.1	0	0.5	1068
3480	2891	135.8	5.57	13.72	7.26	655.6	293.8	0.31	1068
3500	2871	134.6	5.56	13.71	7.26	653.7	293.3	0.30	1065
3600	2771	128.7	5.51	13.67	7.27	644.0	290.7	0.30	1052
3700	2671	123.0	5.46	13.60	7.23	627.9	285.5	0.30	1041
3800	2571	117.3	5.41	13.48	7.19	609.5	279.4	0.30	1031
3900	2471	111.8	5.36	13.36	7.14	591.7	273.4	0.30	1023
4000	2371	106.4	5.31	13.25	7.10	574.4	267.5	0.30	1016
4100	2271	101.0	5.26	13.13	7.06	557.5	261.7	0.30	1010
4200	2171	95.8	5.21	13.02	7.01	540.9	255.9	0.30	1005
4300	2071	90.6	5.16	12.90	6.97	524.6	250.2	0.29	1002
4400	1971	85.4	5.11	12.78	6.92	508.5	244.5	0.29	999
4500	1871	80.4	5.05	12.67	6.87	492.5	238.8	0.29	996
4600	1771	75.4	5.00	12.54	6.83	476.6	233.1	0.29	995
4700	1671	70.4	4.95	12.42	6.77	460.7	227.3	0.29	994
4800	1571	65.5	4.90	12.29	6.72	444.8	221.5	0.29	993
4900	1471	60.7	4.84	12.16	6.67	428.8	215.7	0.28	993
5000	1371	55.9	4.79	12.02	6.62	412.8	209.8	0.28	993

5100	1271	51.2	4.73	11.88	6.56	396.6	203.9	0.28	994
5200	1171	46.5	4.68	11.73	6.50	380.3	197.9	0.28	995
5300	1071	41.9	4.62	11.58	6.44	363.8	191.8	0.28	996
5400	971	37.3	4.56	11.41	6.38	347.1	185.6	0.27	997
5500	871	32.8	4.50	11.24	6.31	330.3	179.4	0.27	999
5600	771	28.2	4.44	11.07	6.24	313.3	173.0	0.27	1000
5650	721	26.1	4.41	10.91	6.09	306.7	153.9	0.27	1001
5701	670	23.8	4.38	10.75	5.95	299.9	154.8	0.28	1001
5701	670	23.8	3.99	10.27	5.57	255.6	123.9	0.29	1001
5736	635	22.4	3.98	10.21	5.54	252.3	122.4	0.29	1001
5771	600	21.0	3.98	10.16	5.52	248.9	121.0	0.29	1000
5821	550	19.1	3.91	9.90	5.37	233.2	112.8	0.29	1000
5871	500	17.1	3.85	9.65	5.22	218.1	105.1	0.29	999
5921	450	15.2	3.79	9.39	5.08	203.7	97.7	0.29	998
5971	400	13.4	3.72	9.13	4.93	189.9	90.6	0.29	997
5971	400	13.4	3.54	8.91	4.77	173.5	80.6	0.30	997
6016	355	11.8	3.52	8.81	4.74	168.2	79.0	0.30	995
6061	310	10.2	3.49	8.73	4.71	163.0	77.3	0.30	994
6106	265	8.7	3.42	8.65	4.68	157.9	75.7	0.29	992
6151	220	7.1	3.44	8.56	4.64	152.9	74.1	0.29	990
6151	220	7.1	3.36	7.99	4.42	127.0	65.6	0.28	990

(cont.)

3

Table 3.9 (*cont.*)

Radius (km)	Depth (km)	Pressure P (Gpa)	Density ρ (g cm^{-3})	Velocity V_P (km s^{-1})	Velocity V_S (km s^{-1})	Bulk modulus K_S (Gpa)	Shear modulus μ (Gpa)	Poisson's ratio σ	Gravity acceleration g (cm s^{-2})
6186	185	5.9	3.36	8.01	4.43	127.8	66.0	0.28	989
6256	115	3.6	3.37	8.03	4.44	128.7	66.5	0.28	988
6291	80	2.5	3.37	8.08	4.47	130.3	67.4	0.28	986
6311	60	1.8	3.38	8.09	4.48	130.7	67.6	0.28	985
6331	40	1.1	3.38	8.11	4.48	131.1	68.0	0.28	984
6346.6	24.4	0.6	3.38	8.11	4.49	131.5	68.2	0.28	984
6346.6	24.4	0.6	2.9	6.8	3.9	75.3	44.1	0.25	984
6356	15	0.3	2.9	6.8	3.9	75.3	44.1	0.25	984
6356	15	0.3	2.6	5.8	3.2	52.0	26.6	0.28	984
6368	3	0.0	2.6	5.8	3.2	52.0	26.6	0.28	983
6368	3	0.0	1.02	1.45	0	0.21	0	0.5	983
6371	0	0.0	1.02	1.45	0	0.21	0	0.5	982

Note: a The Preliminary Reference Earth Model is based on the Earth's free oscillation periods, mass and moment of intertia, as well as seismic wave velocities. It is for an isotropic Earth.

Sources: Dziewonski, A. M. and Anderson, D. L. (1981). *Physics of the Earth and Planetary Interiors*, **25**, 297–356. Poirier, J.-P. (2000). *Introduction to the Physics of the Earth's Interior*, 2nd edn. Cambridge: Cambridge University Press.

3

Table 3.10 Properties of main mantle mineral phases

Mineral	Symbol	Formula	Crystal structure	Density[a] (g cm^{-3})	Bulk modulus[a] (GPa)	Start depth (km)	End depth (km)	End depth pressure (GPa)	Transformation or reaction with increasing depth[b]
Olivine		$(Mg,Fe)_2SiO_4$	Orthorhombic	3.2	129	Top	410	~13.5	Olivine to wadsleyite
Majorite		$(Mg,Fe)SiO_3$	Cubic	3.5	175	Top	~740	~26	
Pyroxene		$Ca(Mg,Fe)Si_2O_6$	Monoclinic	3.3	113	Top	~400?	~13?	Pyroxene to Ca-Pv
Ca-perovskite	Ca-Pv	$CaSiO_3$	Cubic (perovskite)	4.2	~240	~600?	CMB[c]	~135	
Wadsleyite		β-$(Mg,Fe)_2SiO_4$	Orthorhombic	3.5	174	410	520	~18	Wadsleyite to ringwoodite
Ringwoodite		γ-$(Mg,Fe)_2SiO_4$	Cubic (spinel)	3.6	184	520	660	~24	Ringwoodite = Mg-Pv+Mw
Ferropericlase	Mw	$(Mg,Fe)O$	Cubic			660?	CMB	~135	
Perovskite	Mg-Pv	$MgSiO_3$	Orthorhombic		246	~660	~2640	~125	Mg-Pv to Mg-PPv
Post-perovskite	Mg-PPv	$MgSiO_3$	Orthorhombic			~2640	CMB	~135	

Notes: a Properties are at 1 bar (10^5 Pa) and 25 °C.
b Transformations etc. are temperature dependent. c CMB = core–mantle boundary.

Sources: Agee, C. B. (1998). Phase transformations and seismic structure in the upper mantle and transition zone. In *Ultrahigh-Pressure Mineralogy: Physics and Chemistry of the Earth's Interior*, ed. R. J. Hemley, *Reviews in Mineralogy*, vol. 37. Washington, DC: Mineralogical Society of America, pp. 165–203. Price, G. D., ed. (2007). *Treatise on Geophysics*, vol. 2, *Mineral Physics*, ed. G. Schubert. Amsterdam: Elsevier. Price, G. D. (2007). Personal communication.

3

Table 3.11 Main discontinuities in the mantle

Depth[a] (km)	Name or symbol	Affinity	Origin[b]	Transformation details
59	Hales	Global?	Anisotropy and/or transformation	
77	Gutenberg	Oceans	Anisotropy	
230	Lehmann	Continents	Anisotropy	
313	X	Subduction	Transformation	Orthopyroxene modification
410		Global	Transformation	Olivine to wadsleyite
520		Global	Transformation	Wadsleyite to ringwoodite
660[c]		Global	Transformation	Ringwoodite to perovskite + periclase
720		Global?	Transformation	Garnet to perovskite
900		Subduction?	Composition?	
1200		Global?	Transformation	?
1700		?	Transformation?	?
2640	D″		Transformation	Perovskite to post-perovskite phase
2870	ULVZs[d]		Melting	

Notes: *a* The seismic discontinuities are not sharp but occur over several kilometres. *b* Many of the discontinuities result from mineral phase transformations at increasing pressure, others from changes in the degree of seismic anisotropy. *c* The 660 km discontinuity in some places appears to be complicated by mineral transformations in addition to that given above – see the source for further details. *d* ULVZs or ultra-low velocity zones occur at the core–mantle boundary and are not evenly distributed.

Source: Stixrude, L. (2007). Properties of rocks and minerals – seismic properties of rocks and minerals, and structure of the Earth. In *Mineral Physics*, ed. G. D. Price, vol. 2, *Treatise on Geophysics*, ed. G. Schubert. Amsterdam: Elsevier, pp. 7–32.

Table 3.12 Radioactive heat production constants

Isotope	Natural abundance[a] (%)	Half life (a)	Energy per atom ($\times 10^{-12}$ J)	Heat production (W kg^{-1})	Heat production[b] (W)
^{238}U	99.27	4.46×10^9	7.41	9.17×10^{-5}	
^{235}U	0.72	7.04×10^8	7.24	5.75×10^{-4}	
U				9.52×10^{-5}	8.3×10^{12}
^{232}Th	100	1.40×10^{10}	6.24	2.56×10^{-5}	8.6×10^{12}
^{40}K	0.0117	1.26×10^9	0.114	2.97×10^{-5}	
K				3.48×10^{-9}	3.6×10^{12}

Notes: *a* Component percentage abundances for each element. See also Table 8.14.
b The total heat production for the element in the silicate component of the Earth based on the element concentrations in Table 2.5b.

Source: Jaupart, C. and Mareschal, J.-C. (2003). Constraints on crustal heat production from heat flow data. In *The Crust*, ed. R. L. Rudnick, vol. 3, *Treatise of Geochemistry*, ed. H. D. Holland and K. K. Turekian. Oxford: Elsevier-Pergamon, pp. 65–84.

Table 3.13 Radioactive heat production in rocks

Rock	Heat (μW m^{-3})
Granite	1.3–7.1
Granodiorite	0.6–2.0
Syenite	2.2–4.4
Basalt	0.2–0.3

Source: Jessop, A. M. (1990). *Thermal Geophysics.* Amsterdam: Elsevier.

Table 3.14 Earth's heat flow and loss

	Flow (mW m^{-2})	Loss (10^{12} W)
Continental	55–65	
Oceanic	93–101	
Global	80–87	41.0–44.2

Note: Radioactivity accounts for about 60% of the Earth's heat output. Data on the energy balance in the Earth's atmosphere can be found in Chapter 5.

Sources: Pollack, H. N., *et al.* (1993). *Reviews of Geophysics,* **31**, 267–280. Stein, C. A. (1995). Heat flow of the Earth. In *Global Earth Physics. A Handbook of Physical Constants*, ed. T. J. Ahrens. Washington, DC: American Geophysical Union, pp. 144–158.

Table 3.15 Global oceanic heat flow with age of lithosphere

Age (Ma)	Area (10^6 km^2)	Observed (mW m^{-2})
0–1	3.5	131 ± 93
0–2	7.1	136 ± 99
0–4	14.2	128 ± 98
4–9	19.7	82 ± 52
9–20	31.8	103 ± 81
20–35	42.6	64 ± 40
35–52	77	60 ± 34
52–65	29.7	62 ± 26
65–80	37.3	61 ± 27
80–95	27.9	59 ± 43
95–110	24.8	57 ± 20
110–125	15.2	51 ± 13
125–140	16.7	50 ± 20
140–160	8.3	49 ± 14
160–180	3.4	48 ± 10

Source: Stein, C. A. and Stein, S. (1994). Constraints on hydrothermal heat flux through the oceanic lithosphere from global heat flow. *Journal of Geophysical Research*, **99(B2)**, 3081–3095.

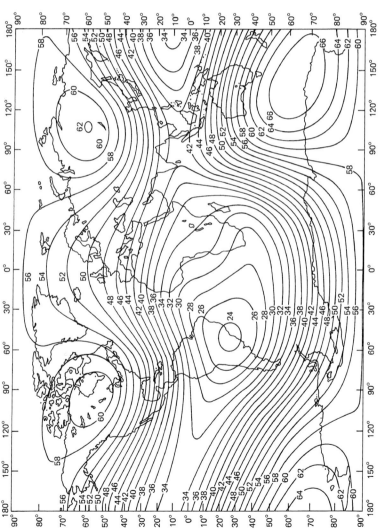

Figure 3.1 Magnitude of the Earth's magnetic field at the present day. (Units: micro tesla, µT)

Source: Turcott, D. L and Schubert, G. (2002). Geodynamics, 2nd edn. Cambridge: Cambridge University Press, with permission. This work also gives charts for the inclination and declination of the magnetic field.

3

Table 3.16 Earthquake magnitude scales

Magnitude name	Symbol	Equation	Limitations
Richter	M_L	$M_L = \log A(\Delta) - \log A_o(\Delta)$ [A, A_o = maximum trace amplitudes (mm) of event wave and of standard wave respectively; Δ = epicentral distance]	Shallow local events only
Body wave	m_b	$m_b = \log(A/T)_{max} + Q(b, \Delta)$ [A = amplitude, T = wave period (s), Q = empirical function of distance, Δ (degrees), and focal depth, b (km) of earthquake.]	Saturates about $m_b = 6.5$
Surface wave	M_S	$M_S = \log(A/T)_{max} + 1.094 \log \Delta + 4.429$ [symbols as above]	Shallow events only. Saturates about $M_S = 8$
Moment magnitude	M_w	$M_w = \frac{2}{3}(\log M_o - 9.1)$ [Seismic moment M_o (units of N m) $= S A \mu$, where S = average amount of slip on fault, A = slip area (km^2) on fault plane and μ = rigidity modulus of faulted rocks. Log $W = 1.5 M_w + 4.8$, where W is the minimum estimate of the strain energy drop as a result of the quake.]	

Notes: The above equations may be adapted to account for local or regional conditions. The scales are not equivalent with respect to earthquake energy output. The values in the equation for the surface wave, M_s, scale have been updated to current usage.

Sources: www.usgs.gov (2006); Doyle, H. (1995). *Seismology*. Chichester: John Wiley and Sons; Udías, A. (1999). *Principles of Seismology*. Cambridge: Cambridge University Press. Abe, K. (1995). Magnitudes and moments of earthquakes. In *Global Earth Physics. A Handbook of Physical Constants*, ed. T. H. Ahrens. Washington, DC: American Geophysical Union, pp. 206–213.

4 Aqueous Earth

Table 4.1 Oceans: mass, volumes and areas

Description		Units
Mass of ocean	1.35×10^{21}	kg
Volume of ocean	1.34×10^{18}	m^3
Volume of surface ocean (0–50 m)	1.81×10^{16}	m^3
Volume of deep ocean (>1200 m)	9.44×10^{17}	m^3
Area of ocean	358×10^{12}	m^2
Annual mean ice-free area of ocean	$\sim 332 \times 10^{12}$	m^2
Area of Atlantic	86.6×10^{12}	m^2
Area of Pacific	166.2×10^{12}	m^2
Area of Indian Ocean	73.43×10^{12}	m^2
Area of Southern Ocean	20.3×10^{12}	m^2
Area of Arctic	9.5×10^{12}	m^2
Area of enclosed seas (Mediterranean, etc)	4.5×10^{12}	m^2
Mean depth of ocean	3690	m
Global mean mixed layer depth	67	m
River input	3.7×10^{13}	$m^3 \, a^{-1}$

Sources: Pilson, M. E. Q. (1998). *An Introduction to the Chemistry of the Sea*. New Jersey: Prentice Hall. Sarmiento, J. L. and Gruber, N. (2006). *Ocean Biogeochemical Dynamics*. Princeton, NJ: Princeton University Press.

Table 4.2 Area and volume of selected ocean basins

Region	Area (10^6 km^2)	Volume (10^6 km^3)
Pacific	166.24	696.19
Asiatic Mediterranean	9.08	11.37
Bering Sea	2.26	3.37
Sea of Okhosk	1.39	1.35
Yellow and East China Seas	1.20	0.33
Sea of Japan	1.01	1.69
Gulf of California	0.15	0.11
Pacific and adjacent seas, total	181.34	714.41
Atlantic	86.56	323.37
American Mediterranean	4.36	9.43
Mediterranean	2.51	3.77
Black Sea	0.46	0.53
Baltic Sea	0.38	0.04
Atlantic and adjacent seas, total	94.27	337.14
Indian	73.43	284.34
Red Sea	0.45	0.24
Persian–Arabian Gulf	0.24	0.02
Indian and adjacent seas, total	74.12	284.61
Arctic	9.49	12.62
Arctic Mediterranean	2.77	1.09
Arctic and adjacent seas, total	12.26	13.70

Sources: Menard, H. W. and Smith, S. M. (1966). *Journal of Geophysical Research*, **71**, 4305–4325. Pilson, M. E. Q. (1998). *An Introduction to the Chemistry of the Sea*. New Jersey: Prentice Hall.

Table 4.3 Areas of structural components of four ocean basins

Component	Pacific	Atlantic	Indian	Arctic	Total[a]
Oceanic rise and ridge	65.11	30.52	22.43	0.51	118.57
Ocean basin	77.95	35.73	36.43	0.00	150.11
Individual volcano	2.13	0.88	0.36	0.00	3.37
Continental shelf and slope	11.30	12.66	6.10	5.87	35.93
Island arc and trench	4.76	0.45	0.26	0.00	5.46
Continental rise	2.69	5.38	4.21	2.27	14.55
Composite volcanic ridge	1.59	0.00	0.41	0.30	2.30
Ridge not known to be volcanic	0.49	0.41	2.57	0.00	3.47
Poorly defined elevation	0.23	0.53	0.68	0.53	1.96
Total	166.24	86.56	73.43	9.48	335.71

Spreading ridges	Total length (km)	5.92×10^4

Notes: Units 10^6 km^2. These data were obtained by hypsometric methods. *a* The totals exclude areas of adjacent seas and gulfs such as the Red Sea, Persian Gulf etc. See Table 4.2.

Sources: Based on: Menard, H. W. and Smith, S. M. (1966). *Journal of Geophysical Research*, **71**, 4305–4325. Sigurdsson, H., ed. (2000). *Encyclopedia of Volcanoes*. San Diego, CA: Academic Press.

Table 4.4 Water volumes in surface-earth reservoirs

Reservoir	Volume (10^6 km^3)	Percent. of total (%)
Oceans	1340	96.83
Surface (0–50 m)	18	1.30
Intermediate (50–1200 m)	378	27.31
Deep (>1200 m)	944	68.21
Ice caps and glaciers	28.4	2.05
Groundwater	15.3	1.11
Lakes	0.125	0.0090
Rivers	0.0017	0.0001
Soil moisture	0.065	0.0047
Atmosphere total	0.0155	0.0011
Terrestrial	0.0045	0.0003
Oceanic	0.011	0.0008
Biosphere	0.002	0.0001

Sources: Oceans: Sarmiento, J. L. and Gruber, W. (2006). *Ocean Biogeochemical Dynamics*. Princeton, NJ: Princeton Univeristy Press. Ice caps: IPCC, Fourth Assessment Report, 2007. Working Group I Report 'The Physical Science Basis', chapter 4. Other freshwater: Berner, E. K. and Berner, R. A. (1966). *Global Environment: Water, Air and Geochemical Cycles*. New Jersey: Prentice Hall.

Table 4.5 Ocean chemistry: element concentrations, residence times and speciation

Z	Element	Conc[A] (ng/kg)	Conc[A] (nmol/kg)	T_{res}[B] (years)	Dist[A]	Species[C]	Equilibrium data[C]
1	H					H^+ (68%), HSO_4^- (29%), HF^0 (3%)	
2	He	7.60×10^0	1.90×10^0		c	He	
3	Li	1.80×10^5	2.59×10^4	2.8×10^6	c	Li^+	
4	Be	2.10×10^{-1}	2.33×10^{-2}	1.0×10^3	s+n	$Be^{2+}/BeOH^+/Be(OH)_2^0$	$pK_1^* = 5.69$; $pK_2^* = 8.25$
5	B	4.50×10^6	4.16×10^5	9.6×10^6	c	$B(OH)_3^0/B(OH)_4^-$	$pK^* = 8.74$
6	C	2.70×10^7	2.25×10^6	8.3×10^4	n	$CO_2/HCO_3^-/CO_3^{2-}$	$pK_1 = 6.00$; $pK_2 = 9.09$
7	N, N_2	8.30×10^6	5.93×10^5		c	NO_3^-, NO_2^-, N_2; NH_4^+/NH_3	$pK = 9.35$
	N, NO_3^-	4.20×10^5	3.00×10^4	3.0×10^3	n		
8	O	2.80×10^6	1.75×10^5		–ve n	O_2	
9	F	1.30×10^6	6.84×10^4	5.0×10^5	c	F^- (50%), MgF^+ (50%)	
10	Ne	1.60×10^2	7.93×10^0		c	Ne	
11	Na	1.08×10^{10}	4.69×10^8	5.5×10^7	c	Na^+ 98%, $NaSO_4^-$ (2%)	
12	Mg	1.28×10^9	5.27×10^7	1.3×10^7	c	Mg_2^+ (90%), $MgSO_4^0$ (10%)	
13	Al	3.00×10^1	1.11×10^0	2.0×10^2	s	$Al(OH)_2^+/Al(OH)_3^0/Al(OH)_4^-$	$pK_3^* = 5.7$; $pK_4^* = 7.7$
14	Si	2.80×10^6	9.97×10^4	2.0×10^4	n	$Si(OH)_4/SiO(OH)_3^-$	$pK = 9.32$

4

15	P	6.20×10^4	2.00×10^3	6.9×10^4	n	$H_2PO_4^-/HPO_4^{2-}/PO_4^{3-}$	$pK_2 = 6.10$; $pK_3 = 8.93$
16	S	8.98×10^8	2.80×10^7	8.7×10^6	c	SO_4^{2-} (33%), $NaSO_4^-$ (35%), $MgSO_4^0$ (20%), HSO_4^-/SO_4^{2-}	$pK \approx 1$
17	Cl	1.94×10^{10}	5.46×10^8	8.7×10^7	c	Cl^- (100%)	
18	Ar	6.20×10^5	1.55×10^4		c	Ar	
19	K	3.99×10^8	1.02×10^7	1.2×10^7	c	K^+ (99%)	
20	Ca	4.12×10^8	1.03×10^7	1.1×10^6	$\approx$c	Ca^{2+} (89%), $CaSO_4^0$ (11%)	
21	Sc	7.00×10^{-1}	1.56×10^{-2}	1.5×10^2	(s+n)	$Sc(OH)_2^+/Sc(OH)_3^0/Sc(OH)_4^-$	$pK_3^* = 6.4$; $pK_4^* = 9.6$
22	Ti	6.50×10^0	1.36×10^{-1}		s+n	$Ti(OH)_3^+/Ti(OH)_4^0$	$pK^* = 2.5$
23	V	2.00×10^3	3.93×10^1	5.0×10^4	$\approx$c	$VO_2(OH)_2^-/VO_3(OH)^{2-}/VO_4^{3-}$	$pK_2 = 7.4$; $pK_3 \leq 13$
24	Cr (VI)	2.10×10^2	4.04×10^0	8.0×10^3	r+n		$pK_3^* = 8.3$; $pK_4^* = 9.1$
	Cr (III)	2.00×10^0	3.85×10^{-2}		r+s	$Cr(OH)_2^+, Cr(OH)_3^0, Cr(OH)_4^-$	
25	Mn	2.00×10^1	3.64×10^{-1}	6.0×10^1	s	Mn^{2+} (72%), $MnCl^+$ (21%)	
26	Fe	3.00×10^1	5.37×10^{-1}	5.0×10^2	s+n	$Fe(OH)_2^+/Fe(OH)_3^0, Fe(OH)_4^-$	$pK_3^* = 6.60$; $pK_4^* = 8.5$
27	Co	1.20×10^0	2.04×10^{-2}	3.4×10^2	s	Co^{2+} (65%), $CoCl^+$ (14%)	
28	Ni	4.80×10^2	8.18×10^0	6.0×10^3	n	Ni^{2+} (53%), $NiCl^+$ (9%)	
29	Cu	1.50×10^2	2.36×10^0	5.0×10^3	s+n	$Cu^{2+}/CuOH^+, Cu^{2+}/CuCO_3^0$	$pK_1^* = 8.11$; $pQ_1 = 6.93$
30	Zn	3.50×10^2	5.35×10^0	5.0×10^4	n	Zn^{2+} (64%), $ZnCl^+$ (16%)	
31	Ga	1.20×10^0	1.72×10^{-2}	5.0×10^2	s+n	$Ga(OH)_2^+/Ga(OH)_3^0/Ga(OH)_4^-$	$pK_3^* = 4.4$; $pK_4^* = 6.0$

(cont.)

4

Table 4.5 (*cont.*)

Z	Element	Conc^A (ng/kg)	Conc^A (nmol/kg)	T_{res} ^{B} (years)	Dist^A	Species^C	Equilibrium data^C
32	Ge	5.50×10^0	7.57×10^{-2}	2.0×10^4	n	$Ge(OH)_4^0/GeO(OH)_3^-/$ $GeO_2(OH)_2^{2-}$	$pK_1 = 8.98$; $pK_2 = 11.8$
33	As (V)	1.20×10^3	1.60×10^1	3.9×10^4	r+n	$H_2AsO_4^-/HAsO_4^{2-}/AsO_4^{3-}/$ $As(OH)_3/As(OH)_4^-$	$pK^* = 9.03$
	As (III)	5.20×10^0	6.94×10^{-2}		r+s		
34	Se (VI)	1.00×10^2	1.27×10^0	2.6×10^4	r+n	SeO_4^{2-} (100%)	$pK_1 = 2.3$; $pK_2 \leq 7.8$
	Se (IV)	5.50×10^1	6.97×10^{-1}		r+n	$H_2SeO_3/HSeO_3^-/SeO_3^{2-}$	
35	Br	6.70×10^7	8.39×10^5	1.3×10^8	c	Br^- (100%)	
36	Kr	3.10×10^2	3.70×10^0		c	Kr	
37	Rb	1.20×10^5	1.40×10^3	3.0×10^6	c	Rb^+ (99%), $RbSO_4^-$ (1%)	
38	Sr	7.80×10^6	8.90×10^4	5.1×10^6	$\approx$c	Sr^{2+} (86%), $SrSO_4^0$ (14%)	
39	Y	1.70×10^1	1.91×10^{-1}	5.1×10^3	n	$Y^{3+}/YCO_3^+/Y(CO_3)_2^-$	$pQ_1 = 6.63$; $pQ_2 = 7.89$
40	Zr	1.50×10^1	1.64×10^{-1}	5.6×10^3	s+n	$Zr(OH)_3^+/Zr(OH)_4^0/Zr(OH)_5^-$	$pK_4^* = 4.6$; $pK_5^* = 5.99$
41	Nb	5.00×10^0	5.38×10^{-2}			$Nb(OH)_4^+/Nb(OH)_5^0/Nb(OH)_6^-$	$pK_5^* \approx -.06$; $pK_6^* = 7.4$
42	Mo	1.00×10^4	1.04×10^2	7.6×10^5	c	$HMoO_4^-/MoO_4^{2-}$	$pK \approx 3.5$
43	Tc					TcO_4^- (100%)	
44	Ru	5.00×10^3	4.95×10^{-5}			$Ru(OH)_n^{4-n}$	pK_n (?)
45	Rh	8.00×10^{-2}	7.77×10^{-4}		n	$RhCl_a(OH)_b^{3-(a+b)}$	pK_n (?)

4

46	Pd	6.00×10^{-2}	5.64×10^{-4}	1.0×10^{4}	n	$PdCl_4^{2-}/PdCl_3OH^{2-}$	$pK^* = 8.7$
47	Ag	2.00×10^{0}	1.85×10^{-2}		n	$AgCl_3^{2-}$ (66%), $AgCl_2^-$ (26%)	
48	Cd	7.00×10^{1}	6.23×10^{-1}	5.0×10^{4}	n	$CdCl^+$ (36%), $CdCl_2^0$ (45%), $CdCl_3^-$ (16%)	
49	In	1.00×10^{-2}	8.71×10^{-5}	1.0×10^{2}	s	$In(OH)_2^+/In(OH)_3^0/In(OH)_4^-$	$pK_3^* = 4.58;\ pK_4^* = 9.37$
50	Sn	5.00×10^{-1}	4.21×10^{-3}	4.0×10^{0}	s	$Sn(OH)_3^+/Sn(OH)_4^0$	$pK^* \approx 1.2$
51	Sb	2.00×10^{2}	1.64×10^{0}	5.7×10^{3}	$\approx$c	$Sb(OH)_5^0/Sb(OH)_6^-$	$pK^* = 2.5$
52	Te (VI)	5.00×10^{-2}	3.92×10^{-4}	1.0×10^{2}	r+s	$Te(OH)_6^0/TeO(OH)_5^-/TeO_2(OH)_4^{2-}$	$pK_1 = 7.35;\ pK_2 \approx 10.2$
	Te (IV)	2.00×10^{-2}	1.57×10^{-4}		r+s	$Te(OH)_4^0/TeO(OH)_3^-/TeO_2(OH)_2^{2-}$	$pK_1 \approx 3.5;\ pK_2 = 8.85$
53	I (V)	5.80×10^{4}	4.57×10^{2}	3.4×10^{5}	$\approx$c	IO_3^- (89%); I^- (100%)	
	I (−1)	4.40×10^{0}	3.47×10^{-2}		(r+s)		
54	Xe	6.60×10^{1}	5.03×10^{-1}		c	Xe	
55	Cs	3.06×10^{2}	2.30×10^{0}	3.3×10^{5}	c	Cs^+ (99%)	
56	Ba	1.50×10^{4}	1.09×10^{2}	1.0×10^{4}	n	Ba^{2+} (86%), $BaSO_4$ (14%)	
57	La	5.60×10^{0}	4.03×10^{-2}	1.0×10^{3}	n	$La^{3+}/LaCO_3^+/La(CO_3)_2^-$	$pQ_1 = 7.38;\ pQ_2 = 8.47$
58	Ce	7.00×10^{-1}	5.00×10^{-3}	1.0×10^{2}	s	$Ce^{3+}/CeCO_3^+/Ce(CO_3)_2^-$	$pQ_1 = 7.05;\ pQ_2 = 8.34$
59	Pr	7.00×10^{-1}	4.97×10^{-3}	5.0×10^{2}	n	$Pr^{3+}/PrCO_3^+/Pr(CO_3)_2^-$	$pQ_1 = 6.88;\ pQ_2 = 8.20$
60	Nd	3.30×10^{0}	2.29×10^{-2}	6.0×10^{2}	n	$Nd^{3+}/NdCO_3^+/Nd(CO_3)_2^-$	$pQ_1 = 6.83;\ pQ_2 = 8.15$
61	Pm					$Pm^{3+}/PmCO_3^+/Pm(CO_3)_2^-$	—
62	Sm	5.70×10^{-1}	3.79×10^{-3}	7.0×10^{2}	n	$Sm^{3+}/SmCO_3^+/Sm(CO_3)_2^-$	$pQ_1 = 6.65;\ pQ_2 = 7.97$

(cont.)

4

Table 4.5 *(cont.)*

Z	Element	Conc[A] (ng/kg)	Conc[A] (nmol/kg)	T_{res}[B] (years)	Dist[A]	Species[C]	Equilibrium data[C]
63	Eu	1.70×10^{-1}	1.12×10^{-3}	6.0×10^2	n	$Eu^{3+}/EuCO_3^+/Eu(CO_3)_2^-$	$pQ_1 = 6.63$; $pQ_2 = 7.89$
64	Gd	9.00×10^{-1}	5.72×10^{-3}	8.0×10^2	n	$Gd^{3+}/GdCO_3^+/Gd(CO_3)_2^-$	$pQ_1 = 6.72$; $pQ_2 = 7.95$
65	Tb	1.70×10^{-1}	1.07×10^{-3}	8.0×10^2	n	$Tb^{3+}/TbCO_3^+/Tb(CO_3)_2^-$	$pQ_1 = 6.65$; $pQ_2 = 7.72$
66	Dy	1.10×10^{0}	6.77×10^{-3}	1.0×10^3	n	$Dy^{3+}/DyCO_3^+/Dy(CO_3)_2^-$	$pQ_1 = 6.55$; $pQ_2 = 7.69$
67	Ho	3.60×10^{-1}	2.18×10^{-3}	2.7×10^3	n	$Ho^{3+}/HoCO_3^+/Ho(CO_3)_2^-$	$pQ_1 = 6.56$; $pQ_2 = 7.59$
68	Er	1.20×10^{0}	7.17×10^{-3}	2.7×10^3	n	$Er^{3+}/ErCO_3^+/Er(CO_3)_2^-$	$pQ_1 = 6.50$; $pQ_2 = 7.53$
69	Tm	2.00×10^{-1}	1.18×10^{-3}	5.0×10^3	n	$Tm^{3+}/TmCO_3^+/Tm(CO_3)_2^-$	$pQ_1 = 6.43$; $pQ_2 = 7.45$
70	Yb	1.20×10^{0}	6.93×10^{-3}	2.2×10^3	n	$Yb^{3+}/YbCO_3^+/Yb(CO_3)_2^-$	$pQ_1 = 6.30$; $pQ_2 = 7.55$
71	Lu	2.30×10^{-1}	1.31×10^{-3}	6.2×10^3	n	$Lu^{3+}/LuCO_3^+/Lu(CO_3)_2^-$	$pQ_1 = 6.36$; $pQ_2 = 7.42$
72	Hf	3.40×10^{0}	1.90×10^{-2}	1.3×10^3		$Hf(OH)_3^+/Hf(OH)_4^0/Hf(OH)_5^-$	$pK_4^* = 4.7$; $pK_5^* = 6.19$
73	Ta	2.50×10^{0}	1.38×10^{-2}			$Ta(OH)_4^+/Ta(OH)_5^0/Ta(OH)_6^-$	$pK_5^* \approx -1$; $pK_6^* = 9.6$
74	W	1.00×10^{1}	5.44×10^{-2}	6.1×10^4	c	HWO_4^-/WO_4^{2-}	$pK \leq 3.5$
75	Re	7.80×10^{0}	4.19×10^{-2}		c	ReO_4^- (100%)	
76	Os	2.00×10^{-3}	1.05×10^{-5}	4.0×10^4		OsO_4^0 (?)	
77	Ir	1.30×10^{-4}	6.76×10^{-7}	2.0×10^3		$IrCl_a(OH)_b^{3-(a+b)}$	pK (?)

4

					Dist	Species	
78	Pt	5.00×10^{-2}	2.56×10^{-4}		(c)	$PtCl_4^{2-}/PtCl_3OH^{2-}$	$pK^* \approx 8.7$
79	Au	2.00×10^{-2}	1.02×10^{-4}	3.8×10^2	(c)	$AuCl_2^-$, $AuCl_3(OH)^-/$ $AuCl_2(OH)_2^-/AuCl(OH)_3^-$	$pK_{22}^* = 6.4$; $pK_{13}^* = 7.4(?)$
80	Hg	1.40×10^{-1}	6.98×10^{-4}	3.5×10^2	(s+n)	$HgCl_4^{2-}$ (88%), $HgCl_3^-$ (12%)	
81	Tl	1.30×10^1	6.36×10^{-2}		$\approx c$	Tl^+ (61%), $TlCl^0$ (37%) $TlCl_4^-/Tl(OH)_3^0$	$pK \approx 7.1$ (?)
82	Pb	2.70×10^0	1.30×10^{-2}	8.0×10^1	s	$PbCl_n^{2-n}/PbCO_3^0$	$pQ = 7.85$
83	Bi	3.00×10^{-2}	1.44×10^{-4}		s	$Bi(OH)_2^+/Bi(OH)_3^0/Bi(OH)_4^-$	$pK_3^* = 4.86$; $pK_4^* \approx 12.7$
88	Ra	1.30×10^{-4}	5.75×10^{-7}		n		
90	Th	2.00×10^{-2}	8.62×10^{-5}	4.5×10^1	s		
92	U	3.20×10^3	1.34×10^1	4.0×10^5	c		

Notes: T_{res} = residence time, $= A/(dA/dt)$, where A is the total amount of the element in suspension or dissolved in seawater; Dist = distribution: c = conservative; n = nutrient; s = scavenged; r = redox controlled.

Sources: A: Nozaki, Y. (1997). A fresh look at element distribution in the North Pacific, http://www.agu.org/eos_elec/97025e.html B: Periodic Table of Elements in the Ocean http://www.mbari.org/chemsensor/pteo.htm C: Byrne, R. H. (2002). Geochemical Transactions, **3**(2), 11–16.

4

Table 4.6 Density variation of seawater with salinity and temperature at one atmosphere

Temperature:	-2	0	5	10	15	20	25	30	35	40
Salinity, S:										
0	-0.33	-0.16	-0.03	-0.30	-0.90	-1.79	-2.95	-4.35	-5.96	-7.78
5	3.78	3.91	3.95	3.61	2.95	2.01	0.81	-0.62	-2.26	-4.09
10	7.86	7.96	7.91	7.50	6.78	5.79	4.56	3.10	1.43	-0.42
15	11.93	11.99	11.86	11.39	10.61	9.58	8.30	6.81	5.12	3.24
20	15.99	16.01	15.81	15.27	14.44	13.36	12.05	10.53	8.81	6.91
25	20.05	20.04	19.76	19.16	18.28	17.15	15.81	14.25	12.51	10.59
30	24.12	24.07	23.71	23.05	22.12	20.95	19.57	17.99	16.22	14.28
32	25.75	25.68	25.30	24.61	23.66	22.48	21.08	19.48	17.70	15.76
33	26.56	26.49	26.09	25.39	24.43	23.24	21.83	20.23	18.45	16.49
34	27.37	27.30	26.88	26.17	25.20	24.00	22.59	20.98	19.19	17.23
35	28.19	28.11	27.68	26.95	25.97	24.76	23.34	21.73	19.93	17.97
36	29.00	28.91	28.47	27.73	26.74	25.53	24.10	22.48	20.68	18.71
38	30.63	30.53	30.06	29.30	28.29	27.05	25.61	23.98	22.17	20.20
40	32.26	32.15	31.65	30.86	29.83	28.58	27.13	25.48	23.66	21.68

Notes: Density in σ_t (i.e. kg m^{-3} − 1000); T in °C; S in ‰

Sources: Modified from Pilson, M. E. Q. (1998). *An Introduction to the Chemistry of the Sea*. New Jersey: Prentice Hall. In turn, from UNESCO *Technical Paper in Marine Science*, (1985). The international system of units (SI) in oceanography.

4

Table 4.7 Selected ocean current transports and locations

Name	Transport (Sv)	Ocean
Agulhas	20–90	South Atlantic
Angola	5	South Atlantic
Antarctic CP	130	Southern
Antarctic Coastal	10	Southern
Antilles	2–7	North Atlantic
Azores	8	North Atlantic
Benguela	7–15	South Atlantic
Brazil	20–70	South Atlantic
Canary	8	North Atlantic
Caribbean	26–33	Caribbean Sea
East Greenland	7–35	North Atlantic
Florida	30	North Atlantic
Guiana	10	North Atlantic
Guinea	3	North Atlantic
Gulf Stream	30–150	North Atlantic
Irminger	2–4	North Atlantic
Labrador	4–8	North Atlantic
Loop Current	30	North Atlantic
Malvinas	10	South Atlantic
Mexican	8–10	Caribbean Sea
North Atlantic	35–40	North Atlantic
North Atlantic Drift	30	North Atlantic
North Brazil	10–30	North Atlantic
North Equatorial	15	North Atlantic
North Equatorial CC	18	North Atlantic
Norwegian	2–4	North Atlantic
South Equatorial	15	South Atlantic
Subtropical CC	4–10	North Atlantic
Weddell Scotia CF	40–90	North Atlantic
Yucatan	23–33	Caribbean Sea
Amazon River	≈0.2	

Notes: CC = counter current, CF = confluence, CP = circumpolar. Flow is in Sverdrups ($1000\,000\ \mathrm{m^3\ s^{-1}}$). The Amazon River value is given for comparison.

Source: University of Miami, CIMAS and NOPP: http://oceancurrents.rsmas.miami.edu/ accessed March, 2008.

Table 4.8 The largest lakes ($>20\,000\,km^2$)

Lake	Location	Area (km^2)
Caspian Sea	Asia–Europe	371 000
Superior	USA/Canada	82 260
Victoria	E Africa	62 940
Huron	USA/Canada	59 580
Michigan	USA	58 020
Tanganyika	E Africa	32 000
Baikal	Russia	31 500
Great Bear	Canada	31 330
Great Slave	Canada	28 570
Erie	USA/Canada	25 710
Winnipeg	Canada	24 390
Malawi/Nyasa	E Africa	22 490

Source: Selected from data in Marsden, H. ed. (2007). *Chambers Book of Facts.* Edinburgh: Chambers Harrap Publishing.

Table 4.9 Major ion concentrations in average rainwater

Ion (mg/litre)	Continental rain	Marine and coastal rain
Na^+	0.2–1	1–5
Mg^{2+}	0.05–0.5	0.4–1.5
K^+	0.1–0.3	0.2–0.6
Ca^{2+}	0.1–3.0	0.2–1.5
NH_4^+	0.1–0.5	0.01–0.05
H^+	pH = 4–6	pH = 5–6
Cl^-	0.2–2	1–10
SO_4^{2-}	1–3	1–3
NO^{3-}	0.4–1.3	0.1–0.5

Source: Modified from Berner, E. K. and Berner, R. A. (1996). *Global Environment: Water, Air and Geochemical Cycles.* London: Prentice Hall.

Table 4.10 Major rivers: discharge, flux and drainage

River	Location	Water (km³ a⁻¹)	Dissolved Solids (Tg a⁻¹)	Suspended Solids (Tg a⁻¹)	Dissolved/ suspended ratio	Drainage area (10⁶ km²)
Amazon	S. America	6300	275	1200	0.23	6.15
Zaire (Congo)	Africa	1250	41	43	0.95	3.82
Orinoco	S. America	1100	32	150	0.21	0.99
Yangtze (Chiang)	Asia (China)	900	247	478	0.53	1.94
Brahmaputra	Asia	603	61	540	0.11	0.58
Mississippi	N. America	580	125	210 (400)	0.60	3.27
Yenisei	Asia (Russia)	560	68	13	5.2	2.58
Lena	Asia (Russia)	525	49	18	2.7	2.49
Mekong	Asia(Vietnam)	470	57	160	0.36	0.79
Ganges	Asia	450	75	520	0.14	0.975
St. Lawrence	N. America	447	45	4	11.3	1.03
Parana	S. America	429	16	79	0.20	2.6
Irrawaddy	Asia (Burma)	428	92	265	0.35	0.43
Mackenzie	N. America	306	64	42	1.5	1.81
Columbia	N. America	251	35	10 (15)	3.5	0.67
Indus	Asia (India)	238	79	59 (250)	1.3	0.975
Red (Hungho)	Asia (Vietnam)	123	?	130	?	0.12
Huanghe (Yellow)	Asia (China)	59	22	1100	0.02	0.77

Notes: Tributaries are excluded; Tg = 1000 000 tonnes; numbers in brackets are pre-dam values.

Sources: Modified from Berner, E. K. and Berner, R. A. (1996). *Global Environment: Water, Air and Geochemical Cycles.* London: Prentice Hall. Water and suspended load data from Milliman, J. D. and Meade, R. H. (1983). *Journal of Geology,* **91**, 1–21; Milliman, J. D. and Syvitski, J. P.M. (1992). *Journal of Geology,* **100**, 525–544.

4

Table 4.11 Chemical composition of average river water

Z	Elem.	World[a]	Amazon[a]	Z	Elem.	World[a]	Amazon[a]
3	Li	1.84	0.91	51	Sb	0.07	0.061
4	Be	0.0089	0.0095	55	Cs	0.011	
5	B	10.2	6.1	56	Ba	23	21
13	Al	32	9.4	57	La	0.120	0.106
21	Sc	1.2		58	Ce	0.262	0.2180
22	Ti	0.489		59	Pr	0.04	0.031
23	V	0.71	0.703	60	Nd	0.152	0.136
24	Cr	0.7	0.717	62	Sm	0.036	0.0349
25	Mn	34	50.73	63	Eu	0.0098	0.0104
26	Fe	66	43	64	Gd	0.04	0.0356
27	Co	0.148	0.1766	65	Tb	0.0055	0.0043
28	Ni	0.801	0.74	66	Dy	0.03	0.033
29	Cu	1.48	1.463	67	Ho	0.0071	0.0064
30	Zn	0.60	0.45	68	Er	0.02	0.018
31	Ga	0.03	0.0174	69	Tm	0.0033	0.0033
32	Ge	0.0068	0.0048	70	Yb	0.0170	0.0159
33	As	0.62		71	Lu	0.0024	0.0020
34	Se	0.07	0.051	72	Hf	0.0059	
37	Rb	1.63	1.49	73	Ta	0.0011	
38	Sr	60.0	25.8	74	W	0.1	
39	Y	0.0400		75	Re	0.0004	0.0002
40	Zr	0.039		76	Os	9.0	4.6
41	Nb	0.0017		82	Pb	0.079	0.064
42	Mo	0.420	0.175	88	Ra	24	9–31
46	Pd	0.028		90	Th	0.041	
48	Cd	0.08	0.1781	92	U	0.372	0.052

Notes: a Units are µg/litre, except for Ra, fg/litre and Os, pg/litre, for dissolved constituents. World is World average. The Amazon River data are given for comparison.

Source: Meybeck, M. (2004). Global occurrence of major elements in rivers. In *Surface and Ground Water, Weathering, and Soils,* ed. J. I. Drever, vol. 5, *Treatise on Geochemistry,* ed. H. D. Holland and K. K. Turekian. Oxford: Elsevier-Pergamon.

Table 4.12 Cryosphere components: areas, volumes and sea-level equivalents

Cryosphere component	Area (10^6 km^2)	Ice volume (10^6 km^3)	Potential sea-level rise (m)
Snow on land (NH)	1.9–45.2	0.0005–0.005	0.001–0.01
Sea ice	19–27	0.019–0.025	0
Glaciers and ice caps	0.51–0.54	0.05–0.13	0.15–0.37
Ice shelves	1.5	0.7	0
Ice sheets	14	27.6	63.9
Greenland	1.7	2.9	7.3
Antarctica	12.3	24.7	56.6
Seasonally frozen ground (NH)	5.9–48.1	0.006–0.065	0
Permafrost (NH)	22.8	0.011–0.037	0.03–0.1

Note: NH = Northern Hemisphere.
Source: The Intergovernmental Panel on Climate Change (2007). *Fourth Assessment Report, The Physical Science Basis*, ch. 4, pp. 337–383 (http://www.ipcc.ch/).

Table 4.13 Fluxes of the surface water cycle

Description	Flux (10^{18} kg a^{-1})
Evaporation from ocean	0.434
Precipitation to ocean	0.398
Evaporation from land	0.071
Precipitation to land	0.107
Land to ocean run off	0.036
Ocean to land vapour transport	0.036

Source: Berner, E. K. and Berner, R. A. (1996). *Global Environment: Water, Air and Geochemical Cycles.* London: Prentice Hall.

Table 4.14 Oxygen and hydrogen isotope values for major meteoric water reservoirs

Reservoir	$\delta^{18}O$ (‰)	δD (‰)
Atmosphere		
Atmosphere over continents	-20 ± 10	-150 ± 80
Atmosphere over oceans	-12 ± 10	
Oceans	0.0	0 ± 5
Freshwater and groundwaters		
Groundwater	-5 ± 15	-40 ± 70
Dilute groundwater	-8 ± 7	-50 ± 60
Brines	0 ± 4	-75 ± 50
Freshwater lakes	-8 ± 7	-50 ± 60
Saline lakes	-2 ± 5	-40 ± 60
Rivers and streams	-8 ± 7	-50 ± 60
Ice caps and glaciers	-30 ± 15	-230 ± 120
Fresh surface water	-8 ± 7	-150 ± 80

Note: $\delta^{18}O = [((^{18}O/^{16}O)_{water}/(^{18}O/^{16}O)_{SMOW}) - 1]1000$, similarly for δD; see also Chapter 8 and Fig. 4.3.

Sources: Criss, R. E. (1999). *Principles of Stable Isotope Distribution.* Oxford: Oxford University Press. Sharp, Z. (2007). *Principles of Stable Isotope Geochemistry.* New Jersey: Pearson Prentice Hall.

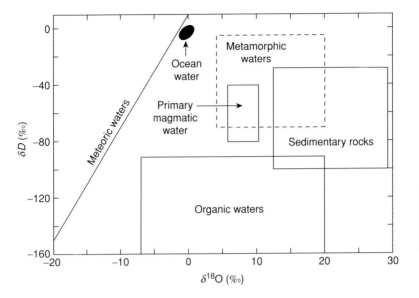

Figure 4.3 Generalised fields of hydrogen and oxygen isotopes for various waters. Data for sedimentary rocks refer to the rocks and not the fluids that may be associated with them. Both δD and δ^{18}O are relative to SMOW which stands for Standard Mean Ocean Water. The meteoric water line has the form δD = 8δ^{18}O + δ$_{excess}$, where δ$_{excess}$ is the deuterium excess. It has a value of 10 in the case of the global meteoric water line. See Chapter 8 for further details. See also Table 4.14 which gives further details about stable isotope values.

Sources: Sharp, Z. (2007). *Principles of Stable Isotope Geochemistry.* New Jersey: Pearson Prentice Hall. Sheppard, S. M. F. (1986). Characterization and isotopic variations in natural waters. In *Stable Isotopes in High Temperature Geological Processes,* ed. J. W. Valley *et al. Reviews in Mineralogy,* vol. 16, Washington DC: Mineralogical Society of America, pp. 165–183.

Table 4.15 Composition of selected seafloor hydrothermal fluids

	East Pacific Rise			Gorda Ridge Escabana Trough	South Cleft
	11°N	13°N	21°N		
T, °C	347	317–380	273–355	108–217	285
pH	3.1–3.7	3.1–3.3	3.3–3.8	5.4	3.2
Element					
Li, μmol/kg	484–884	591–688	891–1322	1286	1108
Na, mmol/kg	290–577	551–596	432–510	560	661
K, mmol/kg	18.7–32.9	27.5–29.8	23.2–25.8	34–40.4	37.3
Rb, μmol/kg	15.0–25.0	14.1–20.0	27.0–33.0	80–105	28
Cs, nmol/kg			202	6.0–7.7	
Ca, mmol/kg	10.6–35.2	44.6–55.0	11.7–20.8	33.4	84.7
Sr, μmol/kg	38–135	168–182	65–97	209	230
Si, mmol/kg	14.3–20.6	17.9–22.0	15.6–19.5	5.6–6.9	22.8
Mn, mmol/kg	0.74–0.93	1.00–2.93	0.67–1.00	0.01–0.02	2.61
Fe, mmol/kg	1.64–6.47	1.45–10.8	0.75–2.43	<0.01	10.3
Cu, μmol/kg					12–17.5
Zn, μmol/kg					330–360
Cl, mmol/kg	338–686	712–760	489–579	668	896
Br, μmol/kg	533–1105	1131–1242	802–929	1179	1580
H_2S, mmol/kg	2.9–8.2	6.6–8.4	3.8–6.0	1.1–1.5	

Source: Von Damm, K. L. (1995). Controls on the chemistry and temporal variation of seafloor hydrothermal fluids. In *Seafloor Hydrothermal Systems: Physical, Chemical, Biological and Geological Interactions*, ed. S. E. Humphris *et al.*, *AGU, Geophysical Monograph*, vol. 91. Washington, DC: American Geophysical Union.

North East Pacific

| Juan de Fuca Ridge | | | | Western Pacific Lau Basin | Mid Atlantic Ridge | |
North Cleft	Axial Volcano	Endeavour Segment	Middle Valley		MARK	TAG
327	299	353	264–276	334	335–350	290–321
3	4.4	4.5	5.13–5.5	2	3.7–3.9	
1434	184	439	370–550	580–745	845	411
682	148	391	315–398	520–615	510	584
40.4	6.98	27.6	13.5–18.7	55–80	23.8	17
25.9		38	22.5–31	60–75	10.7	10
178		364	1000–1400	280–370	179	100
72.9	10.2	42.9	40–81	28–41	9.9–10.5	26
224	46	153	162–257	105–135	51	99
19.9	13.5	17	9.7–10.6	14	18.3	22
1.19	142	194	0.06–0.08	5.8–7.1	0.49	1
2.96	12	533	0.01–0.02	1.2–2.9	1.8–2.2	1.64
7.6	0.4	9	0.3–1.3	15–35		
250	2.2	32	0.7–1.7	1200–3100		
875	176	505	412–578	650–800	559	659
1295	250	895	770–1070	983–1180	847	
3.65		2.9	3		5.9	

4

Table 4.16 Typical compositions of selected deep-sea sediments

Component %	Calcareous	Lithogenous clay	Siliceous
SiO_2	26.96	55.34	63.91
TiO_2	0.38	0.84	0.65
Al_2O_3	7.97	17.84	13.30
Fe_2O_3	3.00	7.04	5.66
FeO	0.87	1.13	0.67
MnO	0.33	0.48	0.50
CaO	0.3	0.93	0.75
MgO	1.29	3.42	1.95
Na_2O	0.8	1.53	0.94
K_2O	1.48	3.26	1.90
P_2O_5	0.15	0.14	0.27
H_2O	3.91	6.54	7.13
$CaCO_3$	50.09	0.79	1.09
$MgCO_3$	2.16	0.83	1.04
Organic C	0.31	0.24	0.22
Organic N	0	0.016	0.016

Sources: Chester, R. (2003). *Marine Geochemistry*, 2nd edn. Malden, MA: Blackwell Publishing. El Wakeel, S. K. and Riley, J. P. (1961). *Geochimica et Cosmochimica Acta*, **25**, 110–146.

Table 4.17 Average compositions of selected marine sediments

Z	Element (μg/g)	Pelagic clay	Fe–Mn nodule	Basal sediment	Ridge sediment	Phosphorite
3	Li	57	80	125		5
4	Be	2.6	2.5	6.7		2.6
5	B	230	300	123	500	16
9	F	1300	200	466		31 000
11	Na%	2.8	1.7	2.56		0.45
12	Mg%	2.1	1.6	2.08		0.18
13	Al%	8.4	2.7	2.73	0.5	0.91
14	Si%	25	7.7	10.8	6.1	5.6
15	P	1500	2500		9000	138 000
16	S	2000	4700			7200
19	K%	2.5	0.7	1.15		0.42
20	Ca%	1	2.3	1.47		31.4
21	Sc	19	10			11
22	Ti	4600	6700		240	640
23	V	120	500		450	100
24	Cr	90	35	15	55	125
25	Mn%	0.67	18.6	6.1	6	0.12
26	Fe%	6.5	12.5	20	18	0.77
27	Co	74	2700	82	105	7
28	Ni	230	6600	460	430	53
29	Cu	250	4500	790	730	75
30	Zn	170	1200	470	380	200
31	Ga	20	10	6.8		4
32	Ge	1.6	0.8	3.3		
33	As, ng/g	20	140		145	23
34	Se	0.2	0.6	2.6		4.6
35	Br		21	58		
37	Rb	110	17	16		
38	Sr	180	830	351		750
39	Y	40	150	128		260
40	Zr	150	560	225		70

(*cont.*)

Table 4.17 (*cont.*)

Z	Element (μg/g)	Pelagic clay	Fe–Mn nodule	Basal sediment	Ridge sediment	Phosphorite
41	Nb	14	50	5.1		10
42	Mo	27	400		30	9
44	Ru, ng/g	0.2	8			
45	Rh, ng/g	0.4	13			
46	Pd, ng/g	6	6		21	
47	Ag	0.11	0.09	0.18	6.2	2
48	Cd	0.42	10	0.4	4	18
49	In	0.08	0.25			
50	Sn	4	2	0.6		3
51	Sb	1	40	17		7
52	Te	1	10			
53	I	28	400			24
55	Cs	6	1			
56	Ba	2300	2300	6230	6000	350
57	La	42	157	98	29	133
58	Ce	101	530	34	8.4	104
59	Pr	10	36	19.3		21
60	Nd	43	158	87	23	98
62	Sm	8.35	35	18.6	5	20
63	Eu	1.85	9	5.4	1.5	6.5
64	Gd	8.3	32	22.6	6	12.8
65	Tb	1.42	5.4			3.2
66	Dy	7.4	31	20.7	7.3	19.2
67	Ho	1.5	7	4.7		4.2
68	Er	4.1	18	12.9	5.6	23.3
69	Tm	0.57	2.3			1.2
70	Yb	3.82	20	13	5.7	13
71	Lu	0.55	1.8	2.2	0.88	2.7
72	Hf	4.1	8	1.6		
73	Ta	1	10	2.1		
74	W	4	100			

4

Table 4.17 (*cont.*)

Z	Element (μg/g)	Pelagic clay	Fe–Mn nodule	Basal sediment	Ridge sediment	Phosphorite
75	Re, ng/g	0.3	1			
76	Os, ng/g	0.14	2			
77	Ir, ng/g	0.4	7		0.8	
78	Pt, ng/g	5	200			
79	Au, ng/g	2	2		16	1.4
80	Hg	0.1	0.15		0.4	0.06
81	Tl	1.8	150	4.8	34	
82	Pb	80	900	100	152	50
83	Bi	0.53	7	0.17		0.06
90	Th	13	30	2.4		6.5
92	U	2.6	5	4.2	22	120

Note: concentrations are in μg/g (ppm), except where stated.

Source: Selected from a compilation by Li, Y.-H. and Schoonmaker, J. E. (2004). Chemical composition and mineralogy of marine sediments. In *Sediments, Diagenesis, and Sedimentary Rocks*, vol. 7, *Treatise on Geochemistry*, ed. H. D. Holland and K. K. Turekian. Oxford: Elsevier-Pergamon.

5 Gaseous Earth

5

Table 5.1 Properties of the atmosphere

Parameter	Value	Units
Average molecular mass	28.97	g mol^{-1}
Total mass	5.3	$\times 10^{18}$ kg
Mass as dry air	5.123	$\times 10^{18}$ kg
Standard surface pressure	101.325	kPa
Average surface density	1.217	kg m^{-3}
Density at 0 °C, 101 325 Pa (dry air)	1.293	kg m^{-3}
Molar volume at 0 °C, 101 325 Pa	0.022 414	m^3 mol^{-1}
Specific gas constant	287	J K^{-1} kg^{-1}
Specific heat at constant P	1004	J K^{-1} kg^{-1}
Specific heat at constant V	717	J K^{-1} kg^{-1}
Thermal conduction at 0 °C, 101 325 Pa	2.4×10^{-2}	J m^{-1} s^{-1} K^{-1}
Refraction index (dry air)	1.000 277	
Speed of sound ('standard, dry, calm air')	343	m s^{-1}
Mean surface temperature	288	K

Sources: Brasseur, G. P. et al., eds. (1999). *Atmospheric Chemistry and Global Change.* Oxford: Oxford University Press. Holton, J. R. et al., eds. (2002). *Encyclopedia of Atmospheric Sciences.* Amsterdam: Elsevier. NASA http://nssdc.gsfc.nasa.gov/planetary updated 19/4/2007, accessed 4/2008.

5

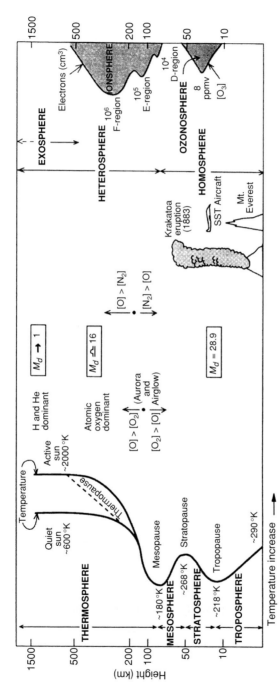

Figure 5.1 Temperature and compositional variation within the atmosphere. The left-hand side shows the temperature variation with height (and its associated terminology); the right-hand side shows the electron density, and the middle section gives data on the variation of the apparent molecular weight of air (M_d) with height.

Source: Hobbs, P. V. (2000). *Introduction to Atmospheric Chemistry.* Cambridge: Cambridge University Press.

5

Table 5.2 Composition of the atmosphere

Element/compound	Formula	Mole fraction[a]	Major source
Nitrogen	N_2	78.084%	Biological
Oxygen	O_2	20.948%	Biological
Argon	Ar	0.934%	Inert
Carbon dioxide[b]	CO_2	384 ppm	Combustion, ocean, biosphere
Neon	Ne	18.18 ppm	Inert
Helium	He	5.24 ppm	Inert
Methane[b]	CH_4	1.78 ppm	Biogenic, anthropogenic
Hydrogen	H_2	0.55 ppm	Biogenic, anthropogenic, photochemical
Nitrous oxide[b]	N_2O	0.31 ppm	Biogenic, anthropogenic
Carbon monoxide	CO	50–200 ppb	Photochemical, anthropogenic
Ozone (troposphere)	O_3	10–500 ppb	Photochemical
Ozone (stratosphere)	O_3	0.5–10 ppm	Photochemical
Non-methane hydrocarbons		5–20 ppb	Biogenic, anthropogenic
Halocarbons (as chlorine)		3.8 ppb	Anthropogenic
Nitrogen oxides	NO_x	10 ppt–1 ppm	Soils, lightning, anthropogenic
Ammonia	NH_3	10 ppt–1 ppb	Biogenic
Hydroxyl radical	OH^-	0.1–10 ppt	Photochemical
Peroxyl	HO_2	0.1–10 ppt	Photochemical
Hydrogen peroxide	H_2O_2	0.1–10 ppb	Photochemical
Formaldehyde	CH_2O	0.1–1 ppb	Photochemical
Sulphur dioxide	SO_2	10 ppt–1 ppb	Photochemical, volcanic, anthropogenic
Dimethyl sulphide	CH_3SCH_3	10–100 ppt	Biogenic
Carbon disulphide	CS_2	1–300 ppt	Biogenic, anthropogenic

(*cont.*)

Table 5.2 (*cont.*)

Element/compound	Formula	Mole fraction[a]	Major source
Carbonyl sulphide	OCS	500 ppt	Biogenic, volcanic, anthropogenic
Hydrogen sulphide	H_2S	5–500 ppt	Biogenic, volcanic
Particulates			
Nitrate	NO_3^-	1 ppt–10 ppb	Photochemical, anthropogenic
Ammonium	NH_4^+	10 ppt–10 ppb	Photochemical, anthropogenic
Sulphate	SO_4^{2-}	10 ppt–10 ppb	Photochemical, anthropogenic

Notes: *a* Mole fraction in dry air. ppm = parts per million; ppb = parts per billion, ppt = parts per trillion. *b* The data for these gases, CO_2, CH_4 and N_2O, are from NOAA (see sources) for the year 2007. The growth rate of global CO_2 concentration has averaged 1.65 ppm per year over the period 1979–2007.

Sources: Brasseur, G. P. et al. eds. (1999). *Atmospheric Chemistry and Global Change.* Oxford: Oxford University Press. NOAA: National Oceanic and Atmospheric Administration web site: http://www.noaanews.noaa.gov Consulted April 2008. Prinn, R. G. (2004). Ozone, hydroxyl radical, and oxidative capacity. In *The Atmosphere*, ed. R. F. Keeling, vol. 4, *Treatise on Geochemistry*, ed. H. D. Holland and K. K. Turekian. Oxford: Elsevier-Pergamon, pp. 1–19.

Table 5.3 Inert-gas isotope composition of the atmosphere

Z	Element/ isotope	Abundance, atomic %	Total abundance		
			Weight (g)	Volume (cm³)	mol
2	Helium		3.707×10^{15}	2.076×10^{19}	9.262×10^{14}
	^{3}He	0.000 14			
	^{4}He	100			
10	Neon		6.484×10^{16}	7.202×10^{19}	3.213×10^{15}
	^{20}Ne	90.50			
	^{21}Ne	0.268			
	^{22}Ne	9.23			
18	Argon		6.594×10^{19}	3.700×10^{22}	1.651×10^{18}
	^{36}Ar	0.3364			
	^{38}Ar	0.0632			
	^{40}Ar	99.60			
36	Krypton		1.688×10^{16}	4.516×10^{18}	2.051×10^{14}
	^{78}Kr	0.3469			
	^{80}Kr	2.2571			
	^{82}Kr	11.523			
	^{83}Kr	11.477			
	^{84}Kr	57.00			
	^{86}Kr	17.398			
54	Xenon		2.019×10^{15}	3.446×10^{17}	1.537×10^{13}
	^{124}Xe	0.0951			
	^{126}Xe	0.0887			
	^{128}Xe	1.919			
	^{129}Xe	26.44			
	^{130}Xe	4.070			
	^{131}Xe	21.22			
	^{132}Xe	26.89			
	^{134}Xe	10.430			
	^{136}Xe	8.857			

5

Sources: Ozima, M. and Podosek, F. A. (2002). *Noble Gas Geochemistry*, 2nd edn. Cambridge: Cambridge University Press. Porcelli, D. and Turekian, K. K. (2004). The history of planetary degassing as recorded by noble gases. In *The Atmosphere*, ed. R. Keeling, vol. 4, *Treatise on Geochemistry*, ed. H. D. Holland and K. K. Turekian. Oxford: Elsevier-Pergamon, pp. 281–318.

Table 5.4 Sources of atmospheric trace gases[a]

Source	SO_2 Tg S a^{-1}	NH_3 Tg N a^{-1}	N_2O Tg N a^{-1}	CH_4 Tg(CH_4) a^{-1}	CO Tg (CO) a^{-1}	NO_x Tg N a^{-1}	NHMC[b] Tg C a^{-1}
Natural:							
Vegetation		5.1			100		400
Wetlands				115			
Animals (wild)		2.5					
Termites				20			
Oceans	25	7.0	3	10	50		50
Soils			6			7	
Lightning						5	
Volcanoes	10						
Other	7.5			15		1.5	
Totals	42.5	14.6	9	160	150	13.5	450
Anthropogenic totals	78	30.4	5.7	360	1000	30.5	110

Notes: a The data are estimated averages for natural sources in the year 2000. Anthropogenic source totals are given for comparison.
b NHMC = non-methane hydrocarbons. See Chapter 6 for tables and figures on element cycles.

Source: Based on data in Wallace, T. M. and Hobbs, P. V. (2006). *Atmospheric Science. An Introductory Survey.* Amsterdam: Elsevier–Academic Press.

5

Table 5.5 Volcanic gas compositions – selected examples

Volcano	Magma type	Year	T °C	log fO₂	H₂O	H₂	CO₂	CO	SO₂	H₂S	S₂	HCL	HF	COS	SO
Convergent plate volcanoes															
Augustine, Alaska, US	Andesite	1979	648	-17.54	97.23	0.381	1.90	0.0035	0.006	0.057		0.365	0.056		
Mt Etna, Italy	Hawaiite	1970	1075	-9.47	46.91	0.50	22.87	0.48	28.70	0.22	0.28				0.06
Mt Etna	Hawaiite	1970	1075	-9.47	49.91	0.54	33.93	0.71	14.69	0.12	0.07				0.02
Mt St. Helens, US	Dacite	1986	802	-14.25	91.58	0.8542	6.942	0.06	0.2089	0.3553	0.0039			0.0008	
Momotombo, Nicaragua	Tholeiite	1985	820	-13.55	97.11	0.70	1.44	0.0096	0.50	0.23	0.0003	2.89	0.259		
Showa-Shinzan, Usu, Japan	Dacite	1960	735	-14.38	99.41	0.17	0.34	0.0004	0.014	0.0008	1.5×10^{-7}	0.042	0.019		
Divergent plate volcanoes															
Erta 'Ale, Ethiopia	Tholeiite	1974	1130	-9.29	76.05	1.57	12.52	0.56	7.59	0.93	0.31	0.42			
Surtsey, Iceland	Alkali basalt	1965	1125	-9.75	87.40	2.86	5.54	0.39	2.72	0.54	0.09				
Hot spot volcano															
Kilauea,[a] East Rift, Hawaiian Is.	Tholeiite	1983	1010	-10.49	79.8	0.9025	3.15	0.0592	14.9	0.622	0.309	0.1	0.019	0.0013	

Notes: Units are mole%; log fO₂ as log bars. The data are for equilibrium compositions. *a* There are many data for Kilauea volcano, including gas emission rates. For example, Gerlach T. M. et al. (2002). *Journal of Geophysical Research*, **107(B9)**, 2189, reported that the CO₂ emission rate over a four-year period of measurement was 8500 t per day. This rate was nearly constant while that for SO₂ showed significant variations.

Source: the data are selected from Symonds, R. B. *et al.* (1994). Volcanic-gas studies: methods, results and applications. In *Volatiles in Magmas*, ed. M. R. Carroll and J. R. Holloway, *Reviews in Mineralogy*, vol. 30. Washington, DC: Mineralogical Society of America.

Table 5.6 Atmospheric aerosol fluxes[a]

Natural source	Average flux (Mt a^{-1}) Ref. A	Flux range $D < 25\mu m$ Ref. B
Primary		
Windblown dust		1000–3000
Mineral dust 0.1–5.0 μm	917	
Mineral dust 5–10 μm	573	
Forest fires		3–150
Sea salt	10 100	1000–10 000
Volcanoes	30	4–10 000
Biological	50	26–50
Secondary		
Sulphates from DMS	12.4	60–110
Sulphates from volcanic SO$_2$	20	10–30
From biogenic VOC	11.2	40–200
From biogenic NO$_x$		10–40

Notes: *a* There is some variation in the data reported for atmospheric aerosol fluxes. Two examples are given here, one of which was published in 2006. Further information can be found in the references, which also give data on anthropogenic sources. These can be significant e.g. 120–180 Mt a^{-1} SO$_2$ from smelters and power plants as reported by reference B. Also, further information on anthropogenic aerosols can be found in the IPCC Fourth Assessment Report: Climate Change 2007.
D = diameter, DMS = dimethyl sulphide, VOC = volatile organic compounds.

Sources: A Seinfeld, J. H. and Pandis, S. N. (2006). *Atmospheric Chemistry and Physics*, 2nd edn. Hoboken, NJ: John Wiley and Sons. Brasseur, G. P. et al. (1999). *Atmospheric Chemistry and Global Change*. New York: Oxford University Press.

Table 5.7 Global biogenic Volatile Organic Carbon (VOC) emissions

Source	VOC flux (Tg a^{-1})
Woods	821
Crops	120
Shrub	194
Ocean	5
Other	9

Source: Guenther, A. et al. (1995). *Journal of Geophysical Research*, **100**, 8873–8892.

Table 5.8 Energy balance: Earth–atmosphere system

	Process	Flux (W m^{-2})
a	Solar radiation: incoming[a]	342
b	Absorbed by atmosphere	67
c	Absorbed by Earth's surface	168
d	Reflected by Earth's surface	30
e	Reflected by clouds	77
	Total	342
f	Outgoing longwave radiation	235
g	Earth surface radiation to atmosphere	390
h	Radiation absorbed by surface from atmosphere	324
i	Heat flux from surface	24
j	Latent heat effects (water vapour)	78
	Balances[2]	
	$b + c + d + e = a$	342
	$f + d + e = a$	342
	$b + g - h + i + j = f$	235
	$c + h = g + i + j$	492

Notes: *a* This solar irradiance is for the top of the atmosphere. The figures indicate that the estimates are for the whole Earth–atmosphere system to be in thermal equilibrium. Data on heat generation and loss within the solid Earth can be found in Chapter 3.

Source: Kiehal, J. T. and Trenberth, K. E. (1997). *Bulletin American Meteorological Society*, **78**, 197–208.

Table 5.9 The Beaufort wind scale

| | Descriptor | Wind speed (m s^{-1}) | |
		Mean	Limits
0	Calm	0	0–0.2
1	Light air	0.8	0.3–1.5
2	Light breeze	2.4	1.6–3.3
3	Gentle breeze	4.3	3.4–5.4
4	Moderate breeze	6.7	5.5–7.9
5	Fresh breeze	9.3	8.0–10.7
6	Strong breeze	12.3	10.8–13.8
7	Near gale	15.5	13.9–17.1
8	Gale	18.9	17.2–20.7
9	Severe gale	22.6	20.8–24.4
10	Storm	26.4	24.5–28.4
11	Violent storm	30.5	28.5–32.6
12	Hurricane		>32.6

Note and source: There are several minor variations of the Beaufort scale in the scientific literature. The version here is based on that of the United Kingdom's Meteorological Office, 20/5/2008: www.metoffice.gov.uk/weather/marine/guide/beaufortscale.html Accessed 21/5/2008.

6 Biological Earth: element cycles

Data on the history of life through the geological record, and stratigraphical ranges of fossil groups can be found in Chapter 7.

6

Table 6.1a Life: divisions and numbers of species

Domain[a]	Includes
Bacteria	Cyanobacteria, Chlamydiae, Proteobacteria etc.
Archaea	Euryarchaeota and Crenarchaeota
Eukarya	See below

Eukarya Domain[a]

	Phylum, group or division	Number of recorded species[c]	Includes[b]
'Protozoa' and Chromista[a]			
	Amoebozoa	Numerous	Slime moulds, certain amoeboids
	Rhizaria	Numerous	Foraminifera, radiolaria
	Alveolata	Numerous	Ciliates, dinoflagellates, animal parasites
	Stramenopila	Numerous	Diatoms, brown algae, kelps
	Haptophyta	Numerous	Coccolithophorids
	Excavata	Numerous	Several parasites, including human ones
Archaeplastida[a]			
(Plantae)	Glaucophyta	13	Microscopic algae
	Rhodophyta	~6000	Red algae, seaweeds, coralline algae
	Chlorophyta	~8000	Green algae
	Hepatophyta	7000	Liverworts
	Anthocerotophyta	100	Hornworts
	Bryophyta	~10 000	Mosses
	Lycopodiophyta	1200	Club mosses
	Equisetophyta	15	Horsetails
	Pteridophyta	11 300	Ferns, ophioglossoid ferns, whisk ferns, rootless fern-like plants
	Cycadophyta	300	Cycads

Table 6.1a (*cont.*)

Phylum, group or division	Number of recorded species[c]	Includes[b]
Ginkgophyta	1	Ginkgo tree only
Pinophyta	630	Conifers (gymnosperms)
Gnetophyta	90	Some woody plants
Magnoliophyta	260 000	Flowering plants (angiosperms)
Fungi[d]		
Basidiomycota	30 000	Mushrooms, toadstools, rusts, yeasts
Ascomycota	~45 000	Truffles, morels, yeasts
Glomeromycota	150	Plant mycorrhiza
Zygomycota	900	Bread moulds
Chytridiomycota	~1 000	Aquatic fungi
Animalia		
Porifera	20 000	Sponges
Placozoa	1	Tablet animal
Ctenophora	166	Sea gooseberries
Cnidaria	9900	Sea anemones, corals, jellyfish
Myxozoa	1300	Small marine parasitic organisms
Acoela	300	Small gutless flatworms
Nemertodermatida	20	Small marine worms
Mesozoa	110	Simple marine parasitic organisms, rhombozoa.
Chordata	~80 000	See Table 6.1c
Hemichordata	106	Pterobranchs, tongue worms
Echinodermata	~7000	Sea lilies, sea urchins, sea cucumbers
Xenoturbellida	2	Worm-like organisms

(cont.)

6

Table 6.1a (*cont.*)

Phylum, group or division	Number of recorded species[c]	Includes[b]
Chaetognatha	179	Marine arrow worms
Loricifera	22	Small unsegmented organisms in marine sediment
Kinorhyncha	150	Mud dragons
Priapulida	18	Priapulid worms
Nematoda	25 000	Roundworms
Nematomorpha	320	Hair worms
Onychophora	178	Velvet worms
Tardigrada	1080	Water bears
Arthropoda	~1.1 million	See Table 6.1b
Platyhelminthes	55 000	Flukes, tapeworms, flatworms
Gastrotricha	710	Very small water organisms, gastrotrichs
Rotifera	2200	Very small water organisms, rotifers
Acanthocephala	1000	Spiny-headed worms
Gnathostomulida	98	Jaw worms
Micrognathozoa	1	Microscopic jaw-bearing organism in spring water
Cycliophora	2	Microscopic parasite on lobster
Bryozoa (= Ectoprocta)	4500	Bryozoans or moss animals
Entoprocta	170	Small aquatic organisms (kamptozoans)
Nemertea	7500	Ribbon worms
Phoronida	20	Tube worms
Brachiopoda	335	Lamp shells
Mollusca	~93 000	Snails, slugs, clams, octopus, squid, chitons

Table 6.1a (*cont.*)

Phylum, group or division	Number of recorded species[c]	Includes[b]
Myzostomida	170	Small marine organisms in association with crinoids
Annelida	16 500	Earthworms, leeches, tube worms, spoon worms
Sipuncula	320	Peanut worms

Notes: *a* Different authorities adopt different classification schemes. For example, many start with six Kingdoms: Eubacteria, Archaebacteria, Protista, Fungi, Plantae and Animalia. New molecular biological research is leading to a much greater understanding of the phylogeny of the eukaryotes and, as a consequence, much fluidity in classificatory schemes and in use of terms, e.g. the Plantae is now included in the Archaeplastida. The 'Protista' in many schemes is no longer formally recognised as a Kingdom but the term, along with 'Protozoa', is still in current usage because of general convenience. For a recent overview of the groupings of the eukaryotes see Baldauf, 2008. The classification and terminology given here for the Eukarya is based mainly on that used by the Catalogue of Life (see below) with some adaptations to take account of recent work – see sources. *b* 'Includes' gives examples, common names or other details of the organisms. *c* These numbers are 'described species', not the estimated total number of species. They are based on data given in the sources below but for some groups, especially the larger ones, there is no firm agreement. The numbers should be taken as guide-line estimates at the present time. *d* Lichens are not included here – they are organisms which are symbiotic between fungi and photosynthetic partners. It is estimated that there are about 17 000 'species' or types of lichens.

Sources: Baldauf, S. L. (2008). An overview of the phylogeny and diversity of eukaryotes. *Journal of Systematics and Evolution*, **46**, 263–273. Bisby, F. A. *et al.*, eds. (2008). Species 2000 & ITIS Catalogue of Life: 2008 Annual Checklist. Digital resource at http://www.catalogueoflife.org/annual-checklist/2008/. Species 2000: Reading, UK. Brusca, R. C. and Brusca, G. J. (2003). *Invertebrates*, 2nd edn. Sunderland, MA: Sinauer Associates. Chapman, A. D. (2005). *Numbers of Living Species in Australia and the World*. Report of the Department of the Environment and Heritage. Canberra, Australia. On line: http://www.deh.gov.au/biodiversity/abrs/publications/other/species-numbers/index.html Giribet, G. *et al.* (2007). A modern look at the animal tree of life. *Zootaxa*, **1668**, 61–79. Jenner, R. A. and Littlewood, D. T. J. (in press). Invertebrate problematica: kinds, causes, and solutions. In *Animal Evolution – Genomes, Fossils and Trees*, ed. M. J. Telford and D. T. J. Littlewood. Oxford: Oxford University Press. Raven, P. H. *et al.* (2005). *Biology*, 7th edn. Boston, MA: McGraw Hill. The Tree of Life Project (ToL) for the fungi: http://tolweb.org/fungi accessed August 2008.

Table 6.1b Arthropoda: classes and numbers of species

Class	Number of recorded species	Includes
Arachnida	98 000	Mites, ticks, harvestmen, scorpions, spiders
Pycnogonida	1300	Sea spiders (pycnogonids)
Merostomata (Xiphosura)	4	Eurypterids, horseshoe crabs
Chilopoda	3460	Centipedes
Diplopoda	11 000	Millipedes
Pauropoda	715	Very small centipede-like animals
Symphyla	200	Garden 'centipedes'
Insecta	950 000	Flies, beetles, moths, butterflies, wasps, bees, praying mantis, dragonflies, grasshoppers, bugs, ants, termites
Entognatha	2000	Springtails
Branchiopoda	~850	Brine shrimp, daphnia
Remipedia	~12	Blind crustaceans (in caves)
Cephalocarida	9	Horseshoe shrimps
Maxillopoda	~9600	Copepods, barnacles
Ostracoda	~5700	Ostracods (sea shrimps)
Malacostraca	~23 000	Crabs, lobsters, shrimps. krill, woodlice

For notes and sources see Table 6.1a.

Table 6.1c Chordata: classes and numbers of species

Class	Number of recorded species	Includes
Ascidiacea	2426	Sea squirts
Thaliacea	74	Marine, free-swimming filter feeders, doliolids, salps
Appendicularia	65	Free-swimming urochordates
Cephalochordata	34	Lancelets
Myxini	70	Hagfish
Cephalaspidomorphi	45	Lampreys
Elasmobranchi	1000	Rays, sharks, skates
Holocephali	40	Ghost sharks, ratfish
Actinopterygii	~29 000	Ray-finned fish
Sarcopterygii	13	Lobe-finned fish
Amphibia	6200	Frogs, newts, salamanders, toads
Reptilia	~13 000	Crocodiles, lizards, snakes, tortoises
Aves	9917	Birds
Mammalia	~7200	Bats, carnivores, primates, whales etc.

For notes and sources see Table 6.1a.

6

Table 6.2 Primary production in natural ecological zones

Ecological zone	Production (Pg a^{-1})
Terrestrial	
Tropical rainforests	17.8
Broadleaf deciduous forest	1.5
Mixed broadleaf and needleleaf forest	3.1
Needleleaf evergreen forest	3.1
Needleleaf deciduous forest	1.4
Savannah	16.8
Perennial grasslands	2.4
Broadleaf shrub	1.0
Tundra	0.8
Desert	0.5
Cultivated land	8.0
Total	56.4
Oceans	48.5
Overall total:	104.9

Note: Primary production is the net gain in carbon by the various ecological zones in unit time. Ocean productivity is dominated by phytoplankton, so most primary production results from photosynthesis. Pg $= 10^{15}$ g or Gt. The estimates here are modelled from satellite and related data.

Source: Field, C. B. *et al.* (1998). *Science*, **281**, 237–240.

Table 6.3 The global nitrogen cycle

Global reservoirs

Reservoir	Amount (TgN)
Atmosphere, N_2	3950×10^6
Sedimentary rocks	999.6×10^6
Ocean N_2	20×10^6
Ocean NO_3^-	570 000
Soil organics	190 000
Land biota	10 000
Marine biota	500

Global fluxes (Tg N a^{-1})		Galloway[a]		Duce[b]
	Year	1890	1990	2008
Emission to atmosphere				
NO_x from land (natural)		6.2	13	14
NO_x anthropogenic		0.6	21	38
NO_x from lightning		5	5	na
NH_3 from land		8.7	43	} 64 (53)
NH_3 from ocean		8	8	
Deposition from atmosphere				
NO_x to land		8	33	na
NO_x to ocean		5	13	23 (17)
Organic fixed N to ocean				20 (16)
NH_3 to land		8	43	na
NH_3 to ocean		12	14	24 (21)
River flux to ocean		5	20	50–80

Notes: NO_x are oxides of N, na = not available. There are numerous estimates of the global N cycle. The two data sets given here are based on: *a* a detailed review by Galloway and *b* a review by Duce *et al.* that focuses on the impact on the open ocean and with added attention paid to anthropogenic fixed N contributions, estimated for the years shown. The numbers in parentheses are the anthropogenic contribution to the total. Duce *et al.* cite uncertainties on the emission data for NO_x of $\pm 30\%$ and $NH_3 \pm 50\%$; on the deposition data of $\pm 40\%$ for NO_x and NH_3 and $\pm 50\%$ for organic fixed N.

Sources: 1. Galloway, J. N. (2004). The global nitrogen cycle. In *Biogeochemistry*, ed. W. H. Schlesinger, vol. 8, *Treatise on Geochemistry*, ed. H. D. Holland and K. K. Turekian. Oxford: Elsevier-Pergamon, pp. 557–583. 2. Duce, R. A. *et al.* (2008). Impacts of atmospheric anthropogenic nitrogen on the open ocean. *Science*, **320**, 893–897.

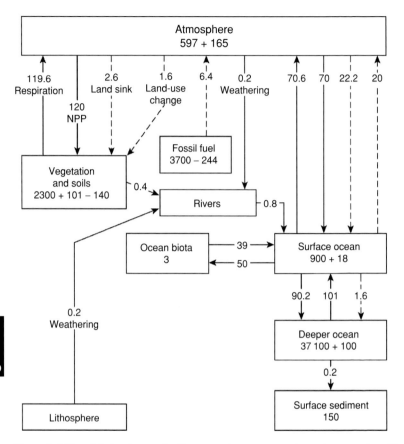

Figure 6.2 The global carbon cycle. The rectangles represent the main reservoirs with the estimated quantities in Pg C. The first number is the natural abundance; the second is the anthropological change. The negative figure (-140) in the 'land biota and soils' reservoir represents the cumulative emissions from anthropological land-use change. The arrows represent fluxes in Pg C a^{-1}; solid arrows are pre-industrial natural fluxes, dashed arrows are anthropogenic ones. Fluxes are estimated to have uncertainties of $\pm20\%$. NPP is natural primary productivity (see also Table 6.2). The cycle is for the 1990s. The sources below should be consulted for more detailed information, especially concerning anthropological effects and also exchanges within the ocean.

Sources: The figure is based on the data given in the IPCC 2007 report: Denman, K. L. *et al.* (2007). Couplings between changes in the climate system and biogeochemistry. In *Climate Change 2007: The Physical Science Basis.* Contribution of Working Group I to the Fourth Assessment Report of the Intergovernmental Panel on Climate Change, ed. S. D. Solomon *et al.* Cambridge: Cambridge University Press. This report uses data from: Field, C. B. and Raupach, M. R., eds. (2004). *The Global Carbon Cycle. Integrating Humans, Climate, and the Natural World.* Washington DC: Island Press, and Sarmiento, J. L. and Gruber, W. (2006). *Ocean Biogeochemical Dynamics.* Princeton, NJ: Princeton University Press. These works can be consulted for further analysis and discussion.

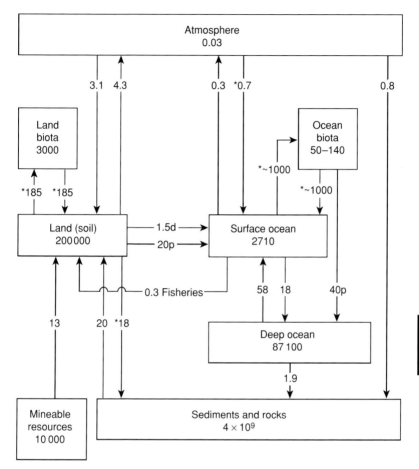

Figure 6.3 The global phosphorus cycle. The rectangles represent the main reservoirs with the estimated quantities as Tg P. The fluxes, represented by arrows, are as Tg P a^{-1}. Flux data followed by 'd' are for dissolved P, those by 'p' are for phosphorus in particulates. Fluxes with their data prefixed by an asterisk have a range of estimates in the scientific literature; the others are consensus or near-consensus figures. Ruttenberg (2004) discusses the ranges in estimates of the various fluxes. Inputs and outputs of certain reservoirs do not exactly balance partly as a result of the uncertainty. The estimated quantity in the 'land biota' reservoir is for the upper 60 cm of soil – this being the location of major interactions involving phosphorus.

Sources: Ruttenberg, K. C. (2004). The global phosphorus cycle. In *Biogeochemistry*, ed. W. H. Schlesinger, vol. 8, *Treatise on Geochemistry*, ed. H. D. Holland and K. K. Turekian. Oxford: Elsevier-Pergamon, pp. 585–643. Jahnke, R. A. (2000). The phosphorus cycle. In *Earth System Science*, ed. M. C. Jacobson et al. San Diego, CA: Academic Press, pp. 360–376.

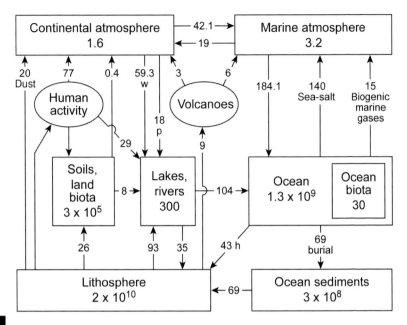

Figure 6.4 The global sulphur cycle. The rectangular areas represent the main reservoirs, with estimated quantities as Tg S. The fluxes, represented by the arrows, are as Tg S a^{-1}. The ovals represent two main processes affecting the sulphur cycle: human activity is principally through the burning of coal; volcanism is mainly through the release of SO$_2$ gas (estimated at about 65%) with the rest being H$_2$S and other gases. The flux from the atmosphere ultimately to lakes and rivers comprises both particulate matter (symbol p) and washout or dry deposit of gases and very small particles (symbol w). The flux of biogenic marine gases to the atmosphere consists mostly of dimethyl sulphide (DMS); which is produced (via the precursor dimethylsulphoniopropianate (DMSP)) by micro-organisms, such as dinoflagellates, in ocean surface waters. This flux is very variable and depends significantly on locality. The flux via rivers to the oceans, as shown, is principally as dissolved sulphate; in addition it is estimated (Brimblecombe, 2004) that about 100 Tg S a^{-1} comes from agriculture and industrial activity and a further 100 Tg S a^{-1} as particulates. The letter h indicates hydrothermal activity leading to deposition of sulphates at mid-oceanic ridges.

Sources: Brimblecombe, P. (2004). The global sulfur cycle. In *Biogeochemistry*, ed. W. H. Schlesinger, vol. 8, *Treatise on Geochemistry*, ed. H. D. Holland and K. K. Turekian. Oxford: Elsevier-Pergamon, pp. 645–682. Charlson, R. J. *et al.* (2000). The sulphur cycle. In *Earth System Science*, ed. M. C. Jacobson *et al.* San Diego, CA: Academic Press, pp. 342–359.

7 Earth history

Table 7.1 Age of the Solar System, Earth and of the oldest rocks and minerals

	Age (Ga)	Location	Type
Solar System	4.567[a]		
Earth	4.51–4.55		
Rocks			
Nuvvuagittuq greenstone belt	4.28[b]	Ungava, Hudson Bay, northern Quebec, Canada	Mafic amphibolite
Acasta gneisses	4.00–4.03[c]	Slave Craton, north-western Canada	Gneisses
Isua greenstone belt	3.7–3.8	Southern west Greenland	Volcanic and sedimentary rocks
Akilia association	3.65–3.70	Godthaab region, west Greenland	Metamorphosed basic, ultrabasic and sedimentary rocks
Minerals			
Zircon	4.404	Jack Hills, Yilgarn Craton, western Australia	Meta conglomerates
Zircon	4.350	Southern Cross, Yilgarn Craton	Quartzites
Diamond-bearing zircon	4.096–4.252	Jack Hills, Yilgarn Craton	Meta conglomerates
Zircon	4.2	Slave Craton, north-western Canada	Gneisses (Acasta)

Notes: *a* This is the earliest age for the Solar System and is based on the work of Amelin, Y., *et al.* (2002). *b* First reported September 2008 by O'Neil, J. *et al.*, but age requires confirmation. *c* The 4 Ga age of the Acasta gneisses has been disputed – see Kamber, B. S. *et al.* (2001) for a discussion. However, more recently, Iizuka, T. *et al.* (2006a) reported on the presence of four generations of tonalite–granite aged between 3.59 and 3.94 Ga, and the age for the Acasta gneisses is considered sound.

Sources: Amelin, Y., *et al.* (2002). Lead isotopic ages of chondrules and calcium–aluminium-rich inclusions. Science, **297**, 1678–1683. Bowring, S. A. and William, I. S. (1999). *Contributions to Mineralogy and Petrology*, **134**, 3–16. Dalrymple, G. B. (2001). The age of the Earth in the twentieth century: a problem (mostly) solved. In *The Age of the Earth: from 4004 BC to AD 2002*, ed. C. L. E. Lewis and S. J. Knell, *Geological Society Special Publications*, vol. 190. London: The Geological Society, pp. 177–203. Kamber, B. S. *et al.* (2001). The oldest rocks on Earth: time constraints and geological controversies. In *The Age of the Earth: from 4004 BC to AD 2002*, ed. C. L. E. Lewis and S. J. Knell, *Geological Society Special Publications*, vol. 190. London: The Geological Society, pp. 205–221. Iizuka, T. *et al.* (2006a). *Precambrian Research*, **153**, 179–208. Iizuka, T. *et al.* (2006b). *Geology*, **34**, 245–248. Menneken, M. *et al.* (2007). *Nature*, **448**, 917–921. O'Neil, J. *et al.* (2008). *Science*, **321**, 1828–1831. Wilde, S. A. *et al.*, (2001). *Nature*, **409**, 175–178. Wyche, S. *et al.*, (2004). *Australian Journal of Earth Science*, **51**, 31–45.

Table 7.2a The geological time scale – Phanerozoic eon summary

Era	System/period	Epoch	Age at base (Ma)
Cenozoic	Quaternary	Holocene	11.7 ka
		Pleistocene	1.806
	Neogene	Pliocene	5.33
		Micoene	23.03
	Paleogene	Oligocene	33.9
		Eocene	55.8
		Paleocene	65.5
Mesozoic	Cretaceous	See Table 7.2c	145.5
	Jurassic	See Table 7.2c	199.6
	Triassic	See Table 7.2c	251.0
Paleozoic	Permian	See Table 7.2d	299.0
	Carboniferous	See Table 7.2d	359.2
	Devonian	See Table 7.2d	416.0
	Silurian	See Table 7.2d	443.7
	Ordovician	See Table 7.2d	488.3
	Cambrian	See Table 7.2d	542.0

Sources for all Tables 7.2: Gradstein, F., Ogg, J. and Smith, A. G. (2004). *A Geologic Time Scale 2004*. Cambridge: Cambridge University Press. International Commission of Stratigraphy (2007). *International Stratigraphic Chart*. Ogg, J. G., Ogg, G. and Gradstein, F. M. (2008). *The Concise Geologic Time Scale*. Cambridge: Cambridge University Press.

Table 7.2b Geological time scale: Cenozoic era

Epoch	Stage	Age at base (Ma)
	Quaternary[a]	
Holocene		0.0117
Pleistocene		1.806
	Neogene	
Pliocene		
Late	Gelasian	2.588
	Piacenzian	3.600
Early	Zanclean	5.332
Miocene		
Late	Messinian	7.246
	Tortonian	11.608
Middle	Serravallian	13.82
	Langhian	15.97
Early	Burdigalian	20.43
	Aquitanian	23.03
	Paleogene	
Oligocene		
Late	Chattian	28.4
Early	Rupelian	33.9
Eocene		
Late	Priabonian	37.2
Middle	Bartonian	40.4
	Lutetian	48.6
Early	Ypresian	55.8
Paleocene		
Late	Thanetian	58.7
Middle	Selandian	~61.1
Early	Danian	65.5

Note: a The status of the Quaternary is not yet finalised.

For sources, see Table 7.2a.

Table 7.2c Geological time scale: Mesozoic era

Epoch	Stage	Age at base (Ma)
	Cretaceous	
Upper	Maastrichtian	70.6
	Campanian	83.5
	Santonian	85.8
	Coniacian	~88.6
	Turonian	93.6
	Cenomanian	99.6
Lower	Albian	112.0
	Aptian	125.0
	Barremian	130.0
	Hauterivian	~133.9
	Valanginian	140.2
	Berriasian	145.5
	Jurassic	
Upper	Tithonian	150.8
	Kimmeridgian	~155.6
	Oxfordian	161.2
Middle	Callovian	164.7
	Bathonian	167.7
	Bajocian	171.6
	Aalenian	175.6
Lower	Toarcian	183.0
	Pliensbachian	189.6
	Sinemurian	196.5
	Hettangian	199.6
	Triassic	
Upper	Rhaetian	203.6
	Norian	216.5
	Carnian	~228.7
Middle	Ladinian	237.0
	Anisian	~245.9
Lower	Olenekian	~249.5
	Induan	251.0

For sources, see Table 7.2a.

Table 7.2d Geological time scale: Paleozoic era

Epoch	Stage	Age at base (Ma)
	Base of Triassic	251
	Permian	
Lopingian	Changhsingian	253.8
	Wuchiapingian	260.4
Guadalupian	Capitanian	265.8
	Wordian	268.0
	Roadian	270.6
Cisuralian	Kungurian	275.6
	Artinskian	284.4
	Sakmarian	294.6
	Asselian	299.0
	Carboniferous	
Pennsylvanian[a]	Gzhelian	303.4
	Kasimovian	307.2
	Moscovian	311.7
	Bashkirian	318.1
Mississippian[a]	Serpukhovian	328.3
	Visean	345.3
	Tournaisian	359.2
	Devonian	
Late	Famennian	374.5
	Frasnian	385.3
Middle	Givetian	391.8
	Eifelian	397.5
Early	Emsian	407.0
	Pragian	411.2
	Lochkovian	416.0

Table 7.2d (*cont.*)

Epoch	Stage	Age at base (Ma)
	Base of Devonian	416.0
	Silurian	
Pridoli		418.7
Ludlow	Ludfordian	421.3
	Gorstian	422.9
Wenlock	Homerian	426.2
	Sheinwoodian	428.2
Llandovery	Telychian	436.0
	Aeronian	439.0
	Rhuddanian	443.7
	Ordovician	
Late	Hirnantian	445.6
	Katian	455.8
	Sandbian	460.9
Middle	Darriwilian	468.1
	Dapingian	471.8
Early	Floian	478.6
	Tremadocian	488.3
	Cambrian[a]	
Furongian	*Stage 10*	~492
	Stage 9	~496
	Paibian	~499
Series 3	Gunzhangian	~503
	Drumian	~506.5
	Stage 5	~510
Series 2	*Stage 4*	~515
	Stage 3	~521
Terreneuvian	*Stage 2*	~528
	Fortunian	~542.0

Note: a The Pennsylvanian and Mississippian are referred to as 'subperiods'. Some divisions in the Cambrian are awaiting ratified definitions.

For sources, see Table 7.2a.

Table 7.2e Geological time scale: the Precambrian

Eon	Era	Period	Age at base (Ma)
Proterozoic	Neoproterozoic	Ediacaran	650
		Cryogenian	850
		Tonian	1000
	Mesoproterozoic	Stenian	1200
		Ectasian	1400
		Calymmian	1600
	Paleoproterozoic	Stratherian	1800
		Orosirian	2050
		Rhyacian	2300
		Siderian	2500
Archean	Neoarchean		2800
	Mesoarchean		3200
	Paleoarchean		3600
	Eoarchean		~4000
	Hadean		~4600

For sources, see Table 7.2a.

Table 7.3a Named polarity chrons

Name[a]	Name	Top (Ma)	Base (Ma)	Duration (Ma)	Type	Span[b]
Brunhes		0	0.781	0.781	Normal	C1n
Matuyama		0.781	2.581	1.800	Reversed	C2r.2r to C1r.1r
	Jaramillo	0.988	1.072	0.084	Normal	C1r.1n
	Cobb Mountain	1.173	1.185	0.012	Normal	C1r.2n
	Olduvai	1.778	1.945	0.167	Normal	C2n
	Reunion	2.128	2.148	0.020	Normal	C2r.1n
Gauss		2.581	3.596	1.015	Normal	C2An
	Kaena	3.032	3.116	0.084	Reversed	C2An.1r
	Mammoth	3.207	3.33	0.123	Reversed	C2An.2r
Gilbert		3.596	6.033	2.437	Reversed	C3r to C2Ar
	Cochiti	4.187	4.300	0.113	Normal	C3n.1n
	Nunivak	4.493	4.631	0.138	Normal	C3n.2n
	Sidufjall	4.799	4.896	0.097	Normal	C3n.3n
	Thvera	4.997	5.235	0.238	Normal	C3n.4n

Notes: *a* The early discovered polarity intervals were given the names of early scientists in geomagnetism. Subsequent work has led to the discovery of shorter intervals within three of the longer ones (Matuyama, Gauss and Gilbert); these were named after their discovery sites. *b* See also Table 7.3b

Source: Gradstein, F., Ogg, J. and Smith, A. G. (2004). *A Geologic Time Scale 2004*. Cambridge: Cambridge University Press.

Table 7.3b The geomagnetic polarity time scale: C-sequence

Polarity chron	Age calibration	
	Base (Ma)	Duration (Ma)
C1	1.778	1.778
C2	2.581	0.803
C2A	4.187	1.606
C3	6.033	1.846
C3A	7.140	1.107
C3B	7.528	0.388
C4	8.769	1.241
C4A	9.779	1.010
C5	12.014	2.235
C5A	13.015	1.001
C5AA	13.369	0.354
C5AB	13.734	0.365
C5AC	14.194	0.460
C5AD	14.784	0.590
C5B	15.974	1.190
C5C	17.533	1.559
C5D	18.056	0.523
C5E	20.040	1.984
C6A	21.083	1.043
C6AA	21.767	0.684
C6B	22.564	0.797
C6C	24.044	1.480
C7	24.915	0.871
C7A	25.295	0.380
C8	26.714	1.419
C9	28.186	1.472
C10	29.451	1.265
C11	30.627	1.176
C12	33.266	2.639

Table 7.3b *(cont.)*

Polarity chron	Age calibration	
	Base (Ma)	Duration (Ma)
C13	34.782	1.516
C15	35.404	0.622
C16	36.512	1.108
C17	38.032	1.520
C18	40.439	2.407
C19	41.590	1.151
C20	45.346	3.756
C21	48.599	3.253
C22	50.730	2.131
C23	52.648	1.918
C24	56.665	4.017
C25	58.379	1.714
C26	61.650	3.271
C27	63.104	1.454
C28	64.432	1.328
C29	65.861	1.429
C30	67.809	1.948
C31	70.961	3.152
C32	73.577	2.616
C33	~84	~10.4
C34	124.61	~40.6

For notes and source, see Table 7.3c.

Table 7.3c The geomagnetic polarity time scale: M-sequence

Polarity chron	Age calibration	
	Base (Ma)	Duration (Ma)
M0	127.24	2.63
M1	127.61	0.37
M3	129.76	2.15
M5	131.19	1.43
M6	131.56	0.37
M7	132.20	0.64
M8	132.83	0.63
M9	133.50	0.67
M10	134.30	0.80
M10N	135.69	1.39
M11	136.90	1.21
M11A	137.60	0.70
M12	138.78	1.18
M12A	139.12	0.34
M13	139.53	0.41
M14	140.66	1.13
M15	141.05	0.39
M16	142.55	1.50
M17	144.04	1.49
M18	144.88	0.84
M19	146.16	1.28
M20	147.77	1.61
M21	148.92	1.15
M22	150.73	1.81
M22A	151.34	0.61
M23	152.26	0.92
M24	153.18	0.92
M24A	153.58	0.40
M24B	154.08	0.50
M25	154.55	0.47

Table 7.3c (*cont.*)

Polarity chron	Age calibration	
	Base (Ma)	Duration (Ma)
M25A	155.05	0.50
M26	155.71	0.66
M27	156.01	0.30
M28	157.26	1.25
M29	157.84	0.58
M30	158.24	0.40
M31	158.63	0.39
M32	158.94	0.31
M33	160.07	1.13
M34	160.73	0.66
M35	160.99	0.26
M36	161.58	0.59
M37	162.17	0.59
M38	163.16	0.99
M39	164.94	1.78
M40	165.55	0.61
M41	na	

Notes: The C-sequence stands for Cenozoic, M-sequence for the Mesozoic. Data are given only for the main numerical subdivisons of the polarity time scale as presently available. Most of these are further subdivided into normal and reversed stages. The details of these can be found in the source reference. The C1 chron continues into the present day. The M0 chron continues to 124.61 Ma.

Source: Based on data in Gradstein, F., Ogg, J. and Smith, A. (2004). *A Geologic Time Scale 2004.* Cambridge: Cambridge University Press.

7

Table 7.4 Main Phanerozoic orogenies

Orogeny	Approx. age[a]	Stratigraphical context	Main location/area
Himalayan	45–0	Eocene to present	Asia
Andean	60–0	Paleocene to present	Western South America
Alpine	65–2	End Cretaceous to end Pliocene	Southern Europe
Laramide	75–45	Late Cretaceous to Middle Eocene	Western North America
Pyrenean	85–25	Late Cretaceous to Late Oligocene	Northern Spain
Sevier	140–65	Cretaceous	Western North America
Nevadan	170–110	Late Jurassic to Early Cretaceous	Western North America
Yanshanian	190–65	Early Jurassic to end Cretaceous	Central and eastern Asia
Kimmerian (Cimmerian)	220–140	Late Triassic to Early Cretaceous	Eastern Europe, western and central Asia
Indosinian	260–200	late Permian to end Triassic	South east Asia
Uralian	300–190	end Carboniferous to Early Jurassic	Eastern Europe, western Siberia
Sonoman	265–240	late Permian to early Triassic	Western North America
Alleghanian[b]	320–260	end Mississippian to late Permian	Eastern North America
Ouachita	330–280	Mississippian to early Permian	Central southern USA
Hercynian[c]	380–300	Late Devonian to end Carboniferous	Western and central Europe

Antler	380–350	Late Devonian to Mississippian	Nevada and NW Utah, USA
Acadian	410–360	Devonian	Eastern North America and western Europe
Caledonian	500–400	Cambrian to Middle Devonian	Scandinavia and Britain
Taconic	470–430	Middle Ordovician to early Silurian	Eastern North America
Delamerian	515–485	Middle Cambrian to Early Ordovician	Southern Australia
Ross	540–490	Cambrian	Antarctica
Penobscotian	550–470	end Ediacaran to mid Ordovician	North America
Grampian	550–470	end Ediacaran to mid Ordovician	Britain
Finnmarkian	550–470	end Ediacaran to mid Ordovician	Norway
Cadomian	640–530	Ediacaran to early Cambrian	North west Europe

Notes: a The time elements associated with many orogenies are a topic of continuing debate. The ages given are approximate. Furthermore, the scope of each orogeny may be a matter of dispute or change – see for example McKerrow, W. S. et al. (2000). The Caledonian Orogeny redefined. Journal of the Geological Society, **157**, 1149–115. b The Alleghanian is sometimes called the Appalachian orogeny. c The name 'Amorican' is sometimes applied to the westerly parts of the Hercynian orogeny, the name 'Variscan' to the remainder.

Sources: Numerous papers and consultations were used for this compilation. These included, as examples: Burchfiel, B. C. and Royden, L. H. (1991). Geology, **19**, 66–69. Encyclopedia Britannica online (2007). Eusden, J. D., Jr., et al. (2000). Journal of Geology, **108**, 219–232. Rogers, J. W. (1993). A History of the Earth. Cambridge: Cambridge University Press. Stanley, S. M. (1993). Exploring Earth and Life Through Time. New York: W. H. Freeman & Co. Tucker, R. D. et al. (2001). American Journal of Science, **301**, 205–260. Wicander, R. and Monroe, J. S. (2004). Historical Geology. Evolution of Earth and Life through Time, 4th edn. Pacific Grove, CA: Brooks/Cole-Thomson Learning.

7

Table 7.5 Main impact structures (>10 km diam.)

Age (Ma)	Name	Location	Latitude	Longitude	Diameter (km)
Cenozoic					
0.9	Zhamanshin	Kazakhstan	N 48° 24′	E 60° 58′	14
1.07	Bosumtwi	Ghana	N 6° 30′	W 1° 25′	10.50
3.5	El'gygtgyn	Russia	N 67° 30′	E 172° 5′	18
<5	Kara-Kul	Tajikistan	N 39° 1′	E 73° 27′	52
15.1	Ries	Germany	N 48° 53′	E 10° 37′	24
35.5	Chesapeake Bay	USA	N37° 17′	W 76° 1′	90
35.7	Popigai	Russia	N 71° 39′	E 111° 11′	100
36.4	Mistastin	Canada	N 55° 53′	W 63° 18′	28
39	Haughton	Canada	N 75° 22′	W 89° 41′	23
~40	Logancha	Russia	N 65° 31′	E 95° 56′	20
42.3	Logoisk	Belarus	N 54° 12′	E 27° 48′	15
49	Kamensk	Russia	N 48° 21′	E 40° 30′	25
50.5	Montagnais	Canada	N 42° 53′	W 64° 13′	45
58	Marquez	USA	N 31° 17′	W 96° 18′	12.7
64.98	Chicxulub	Mexico	N 21° 20′	W 89° 30′	170
65.2	Boltysh	Ukraine	N 48° 45′	E 32° 10′	24
Mesozoic					
<70	Vargeão Dome	Brazil	S 26° 50′	W 52° 7′	12
70.3	Kara	Russia	N 69° 6′	E 64° 9′	65
73.3	Lappajärvi	Finland	N 63° 12′	E 23° 42′	23
73.8	Manson	USA	N 42° 35′	W 94° 33′	35
89	Dellen	Sweden	N 61° 48′	E 16° 48′	19
91	Steen River	Canada	N 59° 30′	W 117° 38′	25
3–95	Avak	USA	N 71° 15′	W 156° 38′	12
<97	Kentland	USA	N 40° 45′	W 87° 24′	13
99	Deep Bay	Canada	N 56° 24′	W 102° 59′	13
<100	Sierra Madera	USA	N 30° 36′	W 102° 55′	13
115	Carswell	Canada	N 58° 27′	W 109° 30′	39

Table 7.5 (*cont.*)

Age (Ma)	Name	Location	Latitude	Longitude	Diameter (km)
<120	Oasis	Libya	N 24° 35′	E 24° 24′	18
128	Tookoonooka	Australia	S 27° 7′	E 142° 50′	55
142	Mjølnir	Norway	N 73° 48′	E 29° 40′	40
142.5	Gosses Bluff	Australia	S 23° 49′	E 132° 19′	22
145.0	Morokweng	South Africa	S 26° 28′	E 23° 32′	70
167	Puchezh-Katunki	Russia	N 56° 58′	E 43° 43′	80
169	Obolon'	Ukraine	N49° 35′	E 32° 55′	20
~200	Wells Creek	USA	N 36° 23′	W 87° 40′	12
214	Manicouagan	Canada	N 51° 23′	W 68° 42′	100
214	Rochechouart	France	N 45° 50′	E 0° 56′	23
220	Saint Martin	Canada	N 51° 47′	W 98° 32′	40
244	Araguainha	Brazil	S 16° 47′	W 52° 59′	40

Paleozoic

Age (Ma)	Name	Location	Latitude	Longitude	Diameter (km)
280	Ternovka	Ukraine	N 48° 08′	E 33° 31′	11
290	Clearwater West	Canada	N 56° 13′	W 74° 30′	36
<300	Serra da Cangalha	Brazil	S 80° 5′	W 46° 52′	12
342	Charlevoix	Canada	N 47° 32′	W 70° 18′	54
<345	Gweni-Fada	Chad	N 17° 25′	E 21° 45′	14
<345	Aorounga	Chad	N 19° 6′	E 19° 15′	12.60
364	Woodleigh	Australia	S 26° 3′	E 114° 39′	>40
377	Siljan	Sweden	N 61° 2′	E 14° 52′	52
380	Kaluga	Russia	N 54° 30′	E 36° 12′	15
<400	Nicholson	Canada	N 62° 40′	W 102° 41′	12.5
~450	Slate Islands	Canada	N 48° 40′	W 87° 0′	30.0
470	Ames	USA	N 36° 15′	W 98° 12′	16.0
<508	Glikson	Australia	S 23° 59′	E 121° 34′	~19
>515	Lawn Hill	Australia	S 18° 40′	E 138° 39′	18

(*cont.*)

Table 7.5 (*cont.*)

Age (Ma)	Name	Location	Latitude	Longitude	Diameter (km)
Precambrian					
>570	Spider	Australia	S 16° 44′	E 126° 5′	13
~590	Acraman	Australia	S 32° 1′	E 135° 27′	90
~600	Beaverhead	USA	N 44° 36′	W 113° 0′	60
646	Strangways	Australia	S 15° 12′	E 133° 35′	25
700	Jänisjärvi	Russia	N 61° 58′	E 30° 55′	14.0
~1620	Amelia Creek	Australia	S 20° 55′	E 134° 50′	~20
1630	Shoemaker	Australia	S 25° 52′	E 120° 53′	30.0
<1800	Keurusselkä	Finland	N 62° 8′	E 24° 36′	~30
1850	Sudbury	Canada	N 46° 36′	W 81° 11′	250
~2000	Yarrabubba	Australia	S 27° 10′	E 118° 50′	30.0
2023	Vredefort	South Africa	S 27° 0′	E 27° 30′	300
~2400	Suavjärvi	Russia	N 63° 7′	E 33° 23′	16

Source: Earth Impact Database, as at July 2007. Further information on the sites can be found at the website: http://www.unb.ca/passc/ImpactDatabase/index.html

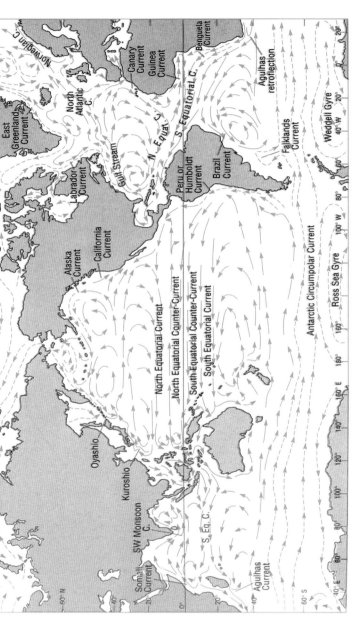

Figure 4.1 Global surface current system – a generalised representation. Cool currents are indicated by dashed arrows, warm currents by solid arrows. The chart is for average conditions for summer months in the Northern Hemisphere; local differences occur in the winter.

Source: Colling, A. *et al.* (2001). *Ocean Circulation*. Oxford: Butterworth-Heinemann in association with the Open University, © The Open University, with permission.

GPCP Sat–Gauge climo (mm/d)

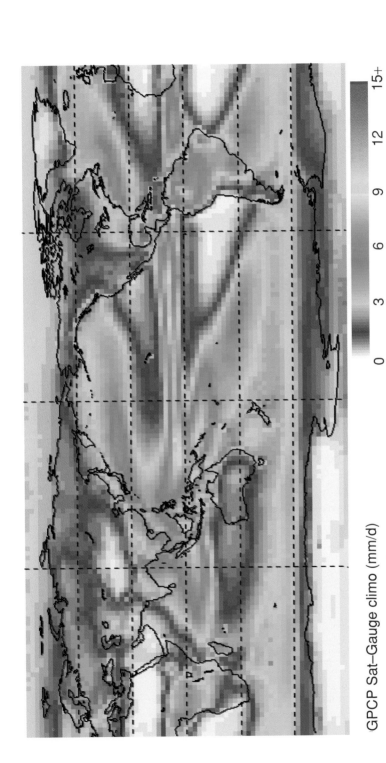

0   3   6   9   12   15+

Figure 4.2 Global rainfall chart. Data are the average daily precipitation rates (mm per day) for the monthly periods from 1979 to 2006 from the Global Precipitation Climatology Project; GPCP Version 2 Monthly Rainrate Climatology Images, for 'All Months'.

Source: http://precip.gsfc.nasa.gov/rain_pages/global_choice.html (Accessed September 2008). This site gives monthly and other charts. GPCP data were provided by NASA/GSFC, visualisation by D. Bolvin (SSAI and NASA/GSFC).

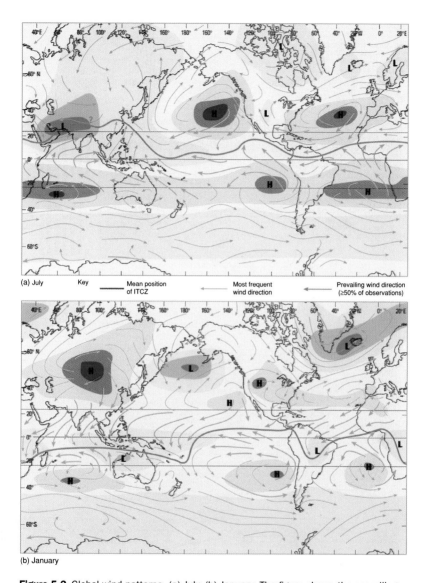

(a) July

Key — Mean position of ITCZ   ← Most frequent wind direction   ← Prevailing wind direction (≥50% of observations)

(b) January

Figure 5.2 Global wind patterns. (a) July, (b) January. The figure shows the prevailing winds and the average position of the Intertropical Convergence Zone (ITCZ, where the trade winds of the Northern Hemisphere and Southern Hemisphere converge), as well as the positions of the main regions of high and low atmospheric pressure.

Source: Colling, A. *et al.* (2001). *Ocean Circulation*. Oxford: Butterworth-Heinemann in association with the Open University, based in part on data in Perry, A. H. and Walker, J. M. (1977). *The Ocean–atmosphere System*. London: Longman, with permission of the publisher.

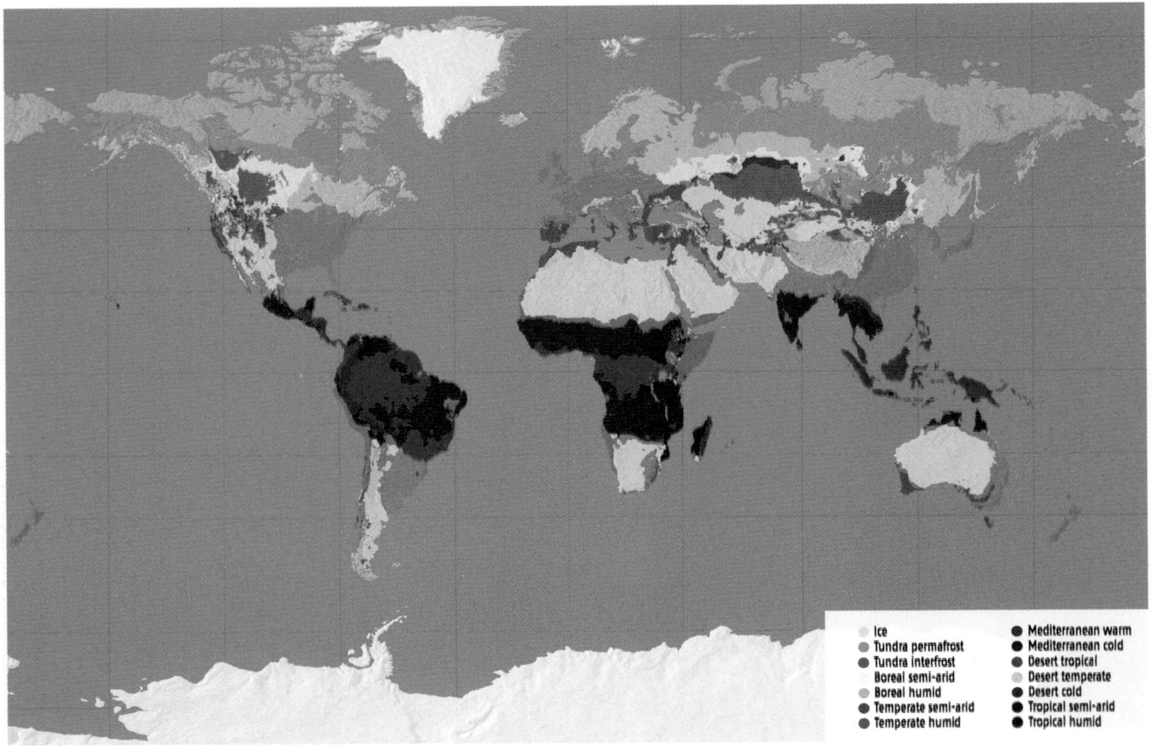

Figure 6.1 World distribution of biomes.

Source: © New Scientist (2003). *The Earth From Inside Out.* With permission.

Ice
Tundra permafrost
Tundra interfrost
Boreal semi-arid
Boreal humid
Temperate semi-arid
Temperate humid

Mediterranean warm
Mediterranean cold
Desert tropical
Desert temperate
Desert cold
Tropical semi-arid
Tropical humid

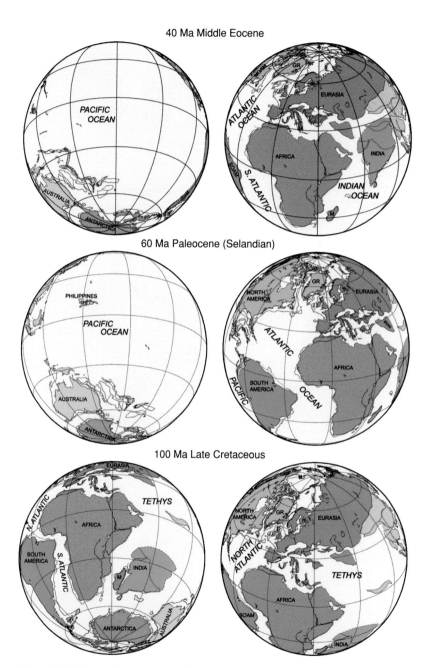

Figure 7.1 Palaeocontinental maps showing the present-day continents in their previous positions. GR, Greenland; M, Madagascar; NOAM, North America and SOAM, South America. Reconstructions were provided by the PLATES Project, Institute for Geophysics, Jackson School of Geosciences University of Texas at Austin.

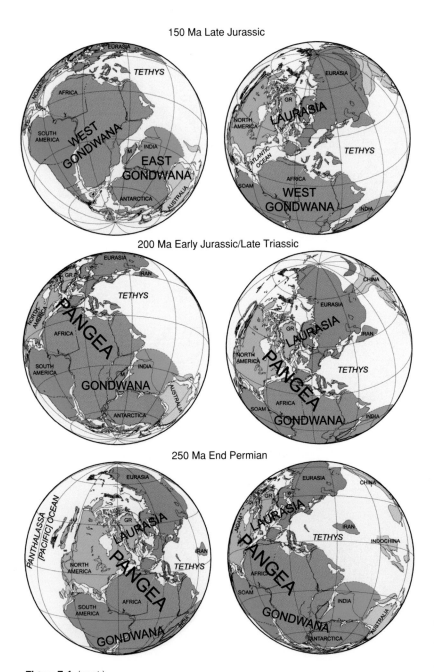

Figure 7.1 (cont.)

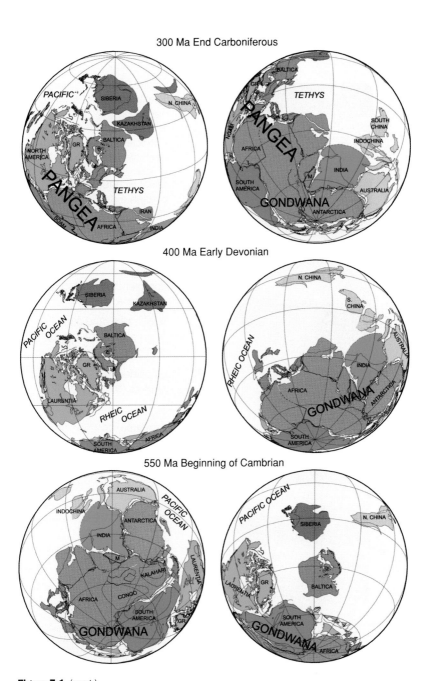

Figure 7.1 (cont.)

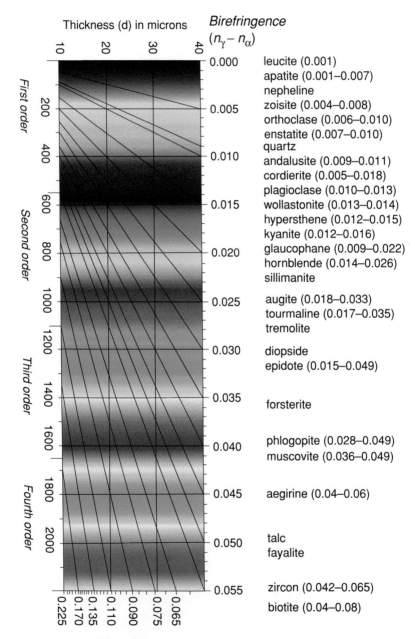

Figure 9.2 Interference colour chart for common minerals. This shows the colour produced by path difference (nm) as a function of mineral thickness (μm) and birefringence. The standard thickness of petrographic thin sections is 30 μm. Specific birefringence values are shown as black lines.

Table 7.6 Main large igneous provinces, last 250 Ma

Province	Age (Ma)	Type[a]	Location	Volume[b] (10^6 km^3)
Columbia River	14.5–16.5	C	Western USA	0.175
Ethiopia–East Africa	15–23; 28–32	C	Eastern Africa	1
North Atlantic	58–61	C	Greenland and W Scotland	1.3×10^6 km^2
	0–56	O	North Atlantic	
Deccan Traps	64–65	C	West India, Seychelles	8.2
Kerguelen Plateau, north	65	O	Southern Ocean	10–15
Kerguelen Plateau, south	115	O	Southern Ocean	
Caribbean	88–90	O	Caribbean Basin and Islands	4.5
Madagascar	88–90	C	Madagascar	4.4
Ontong Java	~122	O	West central Pacific Ocean	44–57
Paraná-Etendeka	129–133	C	Eastern South America	2.5
Karoo-Ferrar Province	179–183	C	Karoo: southern Africa; Ferrar: Antarctica	3.0
Central Atlantic Magmatic Province	191–204	C	NW Africa, SE USA and NE South America	7×10^6 km^2
Siberian Traps	248–250	C	Siberia, Russia	4.0

Note: a O = Oceanic, C = Continental. b Volumes in most cases are only approximate. Areas only are given for N. Atlantic and Central Atlantic Provinces.

Sources: Condie, K. C. (2001). *Mantle Plumes and their Record in Earth History*. Cambridge: Cambridge University Press. Saunders, A. D. (2005). Large igneous provinces: origin and environmental consequences. *Elements*, **1**, 259–263. Saunders, A. D. (2007). Personal communication. Storey, M. *et al.* (2007). *Science*. **316**, 587–589. Web page: http://www.largeigneousprovinces.org (last updated Jan 2, 2004; accessed Sept. 2007).

Table 7.7 Selected major volcanic events, last 600 years[a]

Volcano	Location	Date	VEI[a]
Cerro Hudson	Southern Chile	1991, Aug.	5+
Mt Pinatubo	Luzon, Philippines	1991, June	6
El Chichón	Mexico	1982, April	5
Mt St. Helens	Washington, USA	1980, May	5
Agung	Lesser Sundra Islands, Indonesia	1963, March	5
Bezymianny	Kamchatka Peninsula, Russia	1956, March	5
Kharimkotan	Kuril Islands	1933, Jan.	5
Cerro Azu	Central Chile	1932, April	5+
Novarupta	Alaska Peninsula	1912, June	6
Ksudach	Kamchatka Peninsula, Russia	1907, March	5
Santa Maria	Guatemala	1902, Oct.	6?
Mt Pelée[b]	Martinique, Caribbean	1902, May on	4
Okataina	New Zealand	1886, June	5
Krakatau	Indonesia	1883, Aug.	6
Askja	NE Iceland	1875, March	5
Shiveluch	Kamchatka Peninsula, Russia	1854, Feb.	5
Cosiguina	Nicaragua	1835, Jan.	5
Galunggung	Java, Indonesia	1822, Oct.	5
Tambora	Lesser Sundra Islands, Indonesia	1815, April	7
Mt St. Helens	Washington, USA	1800, Jan. on	5
Shikotsu	Hokkaido, Japan	1739, Aug.	5
Fuji	Honshu, Japan	1707, Dec.	5
Tongkoko	Sulawesi, Indonesia	1680	5?
Gamkonora	Halmahera, Indonesia	1673, May	5?
Shikotsu	Hokkaido, Japan	1667, Sept.	5
Usu	Hokkaido, Japan	1663, Aug.	5
Long Island	NE of New Guinea	~1660	6
Parker	Philippines–Mindanao	1641, Jan.	5?
Komaga-Take	Hokkaido, Japan	1640, July	5
Vesuvius	Italy	1631, Dec.	5?
Furnas	Azores and Madeira	1630, Sept.	5
Huaynaputina	Peru	1600, Feb.	6
Raung	Java, Indonesia	1593	5?

Table 7.7 *(cont.)*

Volcano	Location	Date	VEI[a]
Kelut	Java, Indonesia	1586	5?
Billy Mitchell	Bougainville	~1580	6
Agua de Pau	Azores and Madeira	1563, June	5?
Mt St. Helens	Washington, USA	1482, Jan. on	5
Mt St. Helens	Washington, USA	1480, Jan. on	5+
Bardarbunga	NE Iceland	1477, March?	5?
Sakura-Jima	Kyushu, Japan	1471, Nov.	5?
Kuwae	Vanuatu	~1452	6
Mt Pinatubo	Luzon, Philippines	~1450	5?
Selected earlier key eruptions			
Hekla	S. Iceland	1104, Oct.	5
Baitoushan	E. China	~1000	7
Taupo	New Zealand	180?	7
Vesuvius	Italy	79[c]	5?
Santorini	Greece	~1610 BC	7?
Crater Lake	Oregon, USA	5677 BC	7
Kurile Lake	Kamchatka Peninsula, Russia	~6440 BC	7
Yellowstone	Wyoming, USA	~640 000 BP	8?

Notes: *a* The table lists volcanoes mostly with a volcanic explosivity index (VEI) of 5 or greater. For details on the volcanic explosivity index scale see Chapter 11. *b* The Mt Pelée eruption of 1902 is the most recent devastating one with 29 000 people killed. (Witham, C. S. (2005). *Journal of Volcanology and Geothermal Research,* **148,** 191–233. This paper gives a database on human mortality and other phenomena arising from volcanic activity during the twentieth century.) *c* The famous 'Pompeii' eruption.

Source: Compiled from data in Smithsonian Institution, National Museum of Natural History, Global Volcanism Program www.volcano.si.edu/world/largeeruptions.cfm as at January 2008.

Table 7.8 Major earthquakes 1900–2007

Date	Latitude	Longitude	Magnitude	Region
11/6/1902	50N	148E	8	Sea of Okhotsk
4/1/1903	20S	175W	8	Tonga
11/8/1903	36.36N	22.97E	8.3	Southern Greece
9/7/1905	49N	99E	8.4	Mongolia
23/7/1905	49N	98E	8.4	Central Mongolia
31/1/1906	1N	81.5W	8.8	Off Ecuador coast
17/8/1906	33S	72W	8.2	Valparaiso, Chile
21/10/1907	38N	69E	8	Afghanistan
12/12/1908	14S	78W	8.2	Off coast, central Peru
15/6/1911	28N	130E	8.1	Ryukyu Islands, Japan
26/5/1914	2S	137E	8	West New Guinea
1/5/1915	47N	155E	8	Kuril Islands
1/5/1917	29S	177W	8	Kermadec Islands, New Zealand
26/6/1917	15S	173W	8.4	Tonga
15/8/1918	5.653N	123.563E	8	Celebes Sea
7/9/1918	45.5N	151.5E	8.2	Kuril Islands
30/4/1919	19.823S	172.215W	8.2	Tonga region
5/6/1920	23.5N	122E	8	Taiwan region
20/9/1920	20S	168E	8	Loyalty Islands
11/11/1922	28.553S	70.755W	8.5	Chile–Argentina border
3/2/1923	54N	161E	8.5	Kamchatka, Russia
14/4/1924	7.023N	125.954E	8.3	Mindanao, Philippines
17/6/1928	16.33N	96.7W	8	Oaxaca, Mexico
10/8/1931	47.1N	89.8E	8	Northern Xinjiang, China
3/6/1932	19.84N	103.99W	8.1	Jalisco, Mexico
2/3/1933	39.22N	144.62E	8.4	Sanriku, Japan
15/1/1934	27.55N	87.09E	8.1	Bihar, India
1/2/1938	5.05S	131.62E	8.5	Banda Sea
10/11/1938	55.33N	158.37W	8.2	Shumagin Islands, Alaska
30/4/1939	10.5S	158.5E	8	Solomon Islands
24/5/1940	10.5S	77W	8.2	Off coast, central Peru
25/11/1941	37.17N	18.96W	8.2	Azores, Cape St.Vincent Ridge

Table 7.8 (*cont.*)

Date	Latitude	Longitude	Magnitude	Region
24/8/1942	15S	76W	8.2	Matto Grosso, Brazil
6/4/1943	30.75S	72W	8.2	Off coast, Coquimbo, Chile
7/12/1944	33.75N	136E	8.1	Tonankai, Japan
27/11/1945	24.5N	63E	8	Off coast, Pakistan
1/4/1946	52.75N	163.5W	8.1	Unimak Islands, Alaska
4/8/1946	19.25N	69W	8	Dominican Republic
20/12/1946	32.5N	134.5E	8.1	Nankaido, Japan
24/1/1948	10.5N	122E	8.2	Panay, Philippines
22/8/1949	53.62N	133.27W	8.1	Queen Charlotte Island, Canada
15/8/1950	28.5N	96.5E	8.6	Assam–Tibet
4/3/1952	42.5N	143E	8.1	Hokkaido, Japan
4/11/1952	52.76N	160.06E	9	Kamchatka, Russia
9/3/1957	51.56N	175.39W	8.6	Andreanof Island, Alaska
4/12/1957	45.15N	99.21E	8.1	Gobi-Altai, Mongolia
6/11/1958	44.33N	148.62E	8.3	Kuril Islands
4/5/1959	53.35N	159.65E	8.2	Kamchatka, Russia
22/5/1960	38.24S	73.05W	9.5	Southern Chile
13/10/1963	44.9N	149.6E	8.5	Kuril Islands
28/3/1964	61.02N	147.65W	9.2	Prince William Sound, Alaska
4/2/1965	51.21N	178.5W	8.7	Rat Islands, Alaska
17/10/1966	10.81S	76.68W	8.1	Off coast, Central Peru
16/5/1968	40.90N	143.35E	8.2	Off coast, Honshu, Japan
11/8/1969	43.48N	147.82E	8.2	Kuril Islands
31/7/1970	1.49S	72.56W	8	Colombia
10/1/1971	3.132S	139.70E	8.1	Papua, Indonesia
3/10/1974	12.254S	77.52W	8.1	Off coast, Central Peru
16/8/1976	6.29N	124.09E	8	Mindanao, Philippines
22/6/1977	22.88S	175.9W	8.1	Tonga region
19/8/1977	11.09S	118.46E	8.3	South of Sumbawa, Indonesia

(*cont.*)

Table 7.8 (*cont.*)

Date	Latitude	Longitude	Magnitude	Region
12/12/1979	1.60N	79.36W	8.1	Ecuador
3/3/1985	33.14S	71.87W	8	Off coast, Valparaiso, Chile
19/9/1985	18.19N	102.53W	8	Michoacan, Mexico
7/5/1986	51.52N	174.78W	8	Andreanof Island, Alaska
23/5/1989	52.34S	160.57E	8.1	Macquarie Island area, Australia
9/6/1994	13.84S	67.55W	8.2	La Paz, Bolivia
4/10/1994	43.77N	147.32E	8.3	Kuril Islands
30/7/1995	23.34S	70.29W	8	Northern Chile
9/10/1995	19.06N	104.21W	8	Jalisco, Mexico
17/2/1996	0.89S	136.95E	8.2	Irian Jaya region, Indonesia
25/3/1998	62.88S	149.53E	8.1	Balleny Islands region
16/11/2000	3.98S	152.17E	8	New Ireland region, Papua New Guinea
23/6/2001	16.26S	73.64W	8.4	Off coast, southern Peru
25/9/2003	41.82N	143.91E	8.3	Hokkaido, Japan
23/12/2004	49.31S	161.35E	8.1	Macquarie Island area, Australia
26/12/2004	3.30N	95.98E	9.1	Off west coast, Northern Sumatra
28/3/2005	2.07N	97.01E	8.6	Northern Sumatra, Indonesia
3/5/2006	20.19S	174.12W	8	Tonga
15/11/2006	46.59N	153.23E	8.3	Kuril Islands
13/1/2007	46.24N	154.52E	8.1	East of Kuril Islands
1/4/2007	8.49S	156.99E	8.1	Solomon Islands
15/8/2007	13.35S	76.51W	8	Off coast, Central Peru
12/9/2007	4.52S	101.37E	8.4	Southern Sumatra, Indonesia

Source: US Geological Survey: NEIC Earthquake Data Base to 18 December 2007.
http://neic.usgs.gov/neis/epic

Table 7.9 Selected life and related events in geological past[a]

Epoch base (Ma)	Epoch	Event[a]
11.8 ka	**Holocene**	
	Pleistocene	
		LA Neanderthals (~27 ka)
		FA Cro-Magnons (~40 ka)
		FA *Homo sapiens* (~200 ka)
		LA *Homo heidelbergensis* (~250 ka)
		FA *Homo neanderthalensis* (~400 ka)
		FA *Homo heidelbergensis* (~600 ka)
1.806		LA *Australopithecus robustus* (~1)
	Pliocene	
	Late	FA *Homo erectus* +*ergaster* (~1.9)
		LA *Australopithecus africanus* (~2)
		FA *Homo habilis* (~2.3)
		FA hominins (*Homo rudolfensis*, ~2.5)
		FA *Australopithecus robustus* (~2)
		FA *Australopithecus africanus* (~3)
		LA *Australopithecus afarensis* (~3)
3.600		Panama Isthmus closes (~3)
	Early	FA Australopithecines (*A. anamensis*, ~4)
5.332		FA *Ardipithecus* (*A. ramidus*, 4.4)
	Miocene	
	Late	FA hominines[b] (*Sahelanthropus & Orrorin*) (6–7)
		Messinian salinity crisis (~6–5)
		Spread of grasses
11.608		Gibraltar Isthmus closes (~9)
15.97	Middle	FA *Hippopotamidae* (~15)
23.03	Early	FA hominoids (19)
	Oligocene	
	Late	Red Sea rifting (~24)
28.4		Drake Passage opens (~28)
33.9	Early	Growth of Antarctic ice sheet

7

(cont.)

Table 7.9 (*cont.*)

Epoch base (Ma)	Epoch	Event[a]
	Eocene	
37.2	Late	Early anthropoids (~36)
48.6	Middle	FA rodents
	Early	FA cetaceans (~50)
		FA horses
		Early Eocene Climatic Optimum (~51 to ~53)
		FA bats (*Icaronycteris*) (~52.5)
		FA artiodactyls
55.8		Paleocene–Eocene Thermal Maximum (~55)
	Paleocene	
58.7	Late	
	Middle	FA modern birds
61.1		FA primates
65.5	Early	FA placental mammals (*Placentalia*)
	Cretaceous	
	Late	Mass extinction
		Deccan Traps (~66)
		Ocean floor anoxia event (~93)
		FA grasses
99.6		Radiation of angiosperms
	Early	FA placental mammals (*Eutheria*)
		FA marsupials
		Ocean floor anoxia event (~120)
		South Atlantic opens (~130)
		India parts from Antarctica (~135)
145.5		FA angiosperms (flowering plants) (~140)
	Jurassic	
	Late	*Archaeopteryx* (earliest bird) (155–150)
161.2		FA rudists
	Middle	Pangea starts to fragment
		FA planktonic foraminifera

Table 7.9 (*cont.*)

Epoch base (Ma)	Epoch	Event[a]
175.6		FA crabs (decapods)
199.6	Early	
	Triassic	
	Late	Mass extinction
		Extinction rhynchosaurs and therapsids
		LA conodonts
		FA teleosts (advanced bony fish)
		FA coccolithophores
		FA dinoflagellates
		FA crocodiles
		FA true mammals (210)
		FA dinosaurs (~225)
228.7		FA pterosaurs
	Middle	Marked diversification of therapsids
245.9		FA rhynchosaurs
251.0	Early	FA ichthyosaurs
	Permian	
	Lopingian	Mass extinction
		Siberian Traps (~251)
260.4		Deep sea anoxic event (duration ~20 Ma)
270.6	Guadalupian	FA therapsids
299.0	Cisuralian	
	Carboniferous	
	Pennsylvanian	FA cycads
		FA pelycosaurs
		FA pulmonates (snails and slugs)
		FA isopods
		FA winged insects

(*cont.*)

Table 7.9 (*cont.*)

Epoch base (Ma)	Epoch	Event[a]
318.1		FA reptiles (possibly little earlier)
	Mississippian	Extinction graptolites
		FA conifers (Pinophyta)
359.2		FA mosses (bryophytes)
	Devonian	
		Mass extinction (~370)
	Late	LA placoderms
		FA amphibians
		Extinction ostracoderms
		FA trees (e.g. *Archaeopteris*)
		FA tetrapods (~380)
385.3		FA seed plants (gymnosperms)
	Middle	FA wingless insects
		FA liverworts (Hepatophyta)
397.5		FA ferns (pteridophytes)
	Early	FA lichen
		FA hexapods (springtails)
416.0		FA ammonoids
	Silurian	
	Pridoli	First land arthropods
		FA arachnids
		FA ray-finned fish (actinopterygians)
		FA lobefin fish (sarcopterygians)
		FA headshield fish (cephalaspids)
418.7		FA lungfish (dipnoi)
422.9	Ludlow	Early land plants
	Wenlock	FA jawed fish (gnathostomes inc. placoderms)
428.2		FA *Cooksonia* land plant
443.7	Llandovery	

Table 7.9 (*cont.*)

Epoch base (Ma)	Epoch	Event[a]
	Ordovician	
460.9	Late	Mass extinction
	Middle	FA spores (from plants)
471.8		FA fungi
488.3	Early	FA jawless fish (agnathans inc.ostracoderms)
	Cambrian	
	Furongian	FA gastropods
~499		FA conodonts
	Middle	FA graptolites
~510		FA cephalopods
	Early	FA trilobites +other arthropods
		FA fish (*Haikouichthys*)
		Chordates present (e.g. tunicates)
542		Oldest calcareous algae (~540)
	Precambrian	
	Era:	
	Neoproterozoic	FA metazoans (~600)
		Ediacaran fauna flourish (630–600)
		Further O_2 increase in atmosphere (~900–550)
	Mesoproterozoic	Oldest multicellular algae (1200)
	Paleoproterozoic	Banded iron formations cease (~1700)
		FA eukaryotic alga (*Grypania spiralis*) (2100)
		Banded iron formations abundant (2250–2000)
		'Great oxidation event' atmosphere (2300–2000)
		Stromatolites abundant (2300)
	Neoarchean	Oxygenic photoautotrophs present (~2700)
	Mesoarchaen	

(cont.)

Table 7.9 (*cont.*)

Epoch base (Ma)	Epoch	Event[a]
	Paleoarchean	First stromatolites (~3300)
		Evidence of methanogenic microbes (~3460)
		Evidence of life[c] (~3500)
	Eoarchean	Cessation of major impact phenomenon (~3900)
		Origin of life (~4000?)

Notes: FA = first appearance, LA = last appearance. *a* The selected events are assigned only to a particular epoch or era; the order within any epoch does not reflect the actual sequence of events. More precise dates are given where appropriate (in Ma except where stated). See also Table 7.10 for information on the stratigraphical ranges of certain fossils. The data are based on fossil occurrences – not on molecular estimates. *b* We follow C. Stringer in abbreviating the subfamily *Homininae* to 'hominine' and the tribe *Hominini* to 'hominin'. *c* Evidence is contested.

Sources: Numerous sources were used including: Bateman, R. M. *et al.* (1998). *Annual Review of Ecology and Systematics*, **29**, 263–292. Benton, M. J. (1999). *BioEssays*, **24**, 1043–1051. Benton, M. J. and Ayala, F. J. (2003). *Science*, **300**, 1698–1700. Briggs, D. E. G. and Crowther, P. R. (2001). *Palaeobiology II*. Oxford: Blackwell Science Ltd. Cowen, R. (2005). *History of Life*. 4th edn. Oxford: Blackwell Publishing. Cracraft, J. and Donoghue, P. J., eds. (2004). *Assembling the Tree of Life*. Oxford: Oxford University Press. Donoghue, P. J. and Smith, M. P. (2004). *Telling the Evolutionary Time: Molecular Clocks and the Fossil Record*. London: Taylor and Francis. Falkowski, P. G. *et al.* (2004). *Science*, **305**, 354–360. Han, T. and Runnegar, B. (1992). *Science*, **257**, 232–235. Harland, W. B. *et al.* (1990). *A Geologic Time Scale 1989*. Cambridge: Cambridge University Press. Heckman, D. S. *et al.* (2001). *Science*, **293**, 1129–1133. Isozaki, Y. (1997). *Science*, **276**, 235–238. Kemp, T. S. (2005). *The Origin and Evolution of Mammals*. Oxford: Oxford University Press. Hernick, L. A. *et al.* (2007). *Review of Palaeobotany and Palynology*, **148**, 154–622. Kemp, T. S. (2006). *Journal of Evolutionary Biology*, **19**, 1231–1247. Rampino, M. R. *et al.* (1992). *Celestial Mechanics and Dynamical Astronomy*, **54**, 143–159. Sarangi, S. *et al.* (2004). *Precambrian Research*, **132**, 107–121. Schneider *et al.* (2004). *Nature*, **428**, 553–557. Schopf, J. F. (2006). *Elements*, **2**, 229–233. Stringer, C. (2008). Personal communication. Stringer, C. and Andrews, P. (2005). *The Complete World of Human Evolution*. London: Thames and Hudson. Thewissen, J. G. M. *et al.* (2007). *Nature*, **450**, 1190–1194. Ueno, Y. *et al.* (2006). *Nature*, **440**, 516–519. Wellman, C. H. *et al.* (2003). *Nature*, **425**, 282–285.

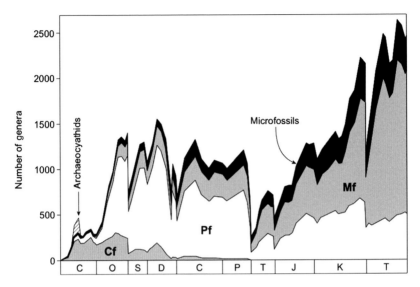

Figure 7.2 Variation in the number of genera through the Palaeozoic according to Sepkoski (1997). Cf, Cambrian faunas, Pf, Palaeozoic faunas, and Mf, Modern faunas. The data do not include taxa known from only one single stratigraphic interval. Figure from: Sepkoski, J. J., Jr. (1997). Biodiversity: past, present and future. *Journal of Paleontology*, **71**, 533–539, with permission of the Paleontological Society.

Table 7.10 Selected fossil stratigraphical ranges according to *The Fossil Record 2*

Kingdom or Phylum[a]	Class[a]	Subclass, superorder or order[a]	Range	
			First	Last
Fungi			Silurian, Wenlock	Holocene
'*Algae*'				
		Prymnesiophyta (coccoliths)	Triassic, Late	Holocene
		Bacillariophyceae (diatoms)	Jurassic, Late	Holocene
		Dinophyceae (dinoflagellates)	Triassic, Late	Holocene
		Acritarcha (*Group*)	Neoproterozoic	Holocene
'*Protozoa*'			Cambrian, Early	Holocene
	Granuloreticulosea (foraminifera)		Cambrian, Early	Holocene

(cont.)

Taxon		
Cnidaria (Coelenterata)	Neoproterozoic	Holocene
Anthozoa	Cambrian, Middle	Holocene
Alcyonaria (Octocorallia)	Triassic, Early	Holocene
Scleractinia	Triassic, Middle	Holocene
Tabulata	Ordovician, Early	Permian, Lopingian
Rugosa	Ordovician, Late	Permian, Lopingian
Mollusca	Cambrian, Early	Holocene
Gastropoda	Cambrian, Furongian	Holocene
Archaeogastropoda	Cambrian, Furongian	Holocene
Apogastropoda	Ordovician, Early	Holocene
Tectibranchia	Jurassic, Early	Holocene
Basommatophora	Carboniferous, Early Mississippian	Holocene
Stylommatophora	Carboniferous, Middle Pennsylvanian	Holocene
Cephalopoda	Cambrian, Furongian	Holocene
Ellesmerocerida	Cambrian, Furongian	Ordovician, Late
Endocerida	Ordovician, Late	Silurian, Wenlock
Actinocerida	Ordovician, Early	Carboniferous, Early Pennsylvanian

7

Table 7.10 (*cont.*)

Kingdom or Phylum[a]	Class[a]	Subclass, superorder or order[a]	Range	
			First	Last
	Cephalopoda (*cont.*)	Orthocerida	Ordovician, Early	Triassic, Late
		Oncocerida	Ordovician, Early	Carboniferous, Mississippian, Middle
		Discosorida	Ordovician, Early	Devonian, Late
		Nautiloida	Devonian, Early	Holocene
		Anarcestida	Devonian, Early	Triassic, Late
		Goniatitida	Devonian, Middle	Triassic, Early
		Ceratitida	Permian, Cisuralian	Triassic, Late
		Ammonoidea	Devonian, Early	Cretaceous, Late
		Belemnitida	Jurassic, Early	Cretaceous, Late
	Bivalvia	Nuculoida	Cambrian, Early	Holocene
		Mytiloida	Cambrian, Early	Holocene
		Arcoida	Cambrian, Middle	Holocene
		Pterioida	Ordovician, Early	Holocene
		Ostreoida	Ordovician, Middle	Holocene
			Ordovician, Early	Holocene

(cont.)

Taxon		
Lucinoida	Ordovician, Early	Holocene
Trigonioida	Silurian, Ludlow	Holocene
Veneroida	Devonian, Early	Holocene
Hippuritoida	Silurian, Llandovery	Paleocene
Pholadomyoida	Ordovician, Late	Holocene
Myoida	Triassic, Middle	Holocene
Arthropoda		
Trilobita	Cambrian, Early	Permian, Lopingian
Redlichiida (inc. Olenellids)	Cambrian, Early	Cambrian, Middle
Agnostida	Cambrian, Early	Ordovician, Late
Corynexochida	Cambrian, Early	Devonian, Late
Lichida	Cambrian, Middle	Devonian, Late
Phacopida	Cambrian, Late	Devonian, Late
Ptychopariida	Cambrian, Middle	Devonian, Late
Asaphida	Cambrian, Middle	Silurian, Ludlow
Proetida	Cambrian, Late	Permian, Lopingian
Crustacea	Cambrian, Early	Holocene
Hexapoda, Insecta	Devonian, Early	Holocene
Coleoptera	Permian, Cisuralian	Holocene
Collembola	Devonian, Early	Holocene

7

Table 7.10 (*cont.*)

Kingdom or Phylum[a]	Class[a]	Subclass, superorder or order[a]	Range First	Range Last
	Hexapoda, Insecta (*cont.*)	Diptera	Permian, Guadalupian	Holocene
		Hemiptera	Permian, Cisuralian	Holocene
		Hymenoptera	Triassic, Late	Holocene
		Lepidoptera	Jurassic, Late	Holocene
		Odonata	Permian, Cisuralian	Holocene
		Orthoptera	Carboniferous, Pennsylvanian	Holocene
Brachiopoda				
	Lingulata	Lingulida	Cambrian, Early	Holocene
		Acrotretida	Cambrian, Early	Holocene
			Cambrian, Early	Holocene
	Articulata		Cambrian, Early	Holocene
		Orthida	Cambrian, Early	Devonian, Frasnian
		Enteletida	Ordovician, Early	Triassic, Early
		Strophomenida	Ordovician, Early	Permian, Lopingian
		Pentamerida	Cambrian, Early	Devonian, Frasnian
		Rhynconellida	Ordovician, Middle	Holocene
		Atrypida	Ordovician, Late	Devonian, Frasnian

(cont.)

Phylum	Order/Class	First appearance	Last appearance
	Spiriferida	Ordovician, Late	Holocene
	Terebratulida	Silurian, Ludlow	Holocene
Echinodermata		Cambrian, Early	Holocene
	Blastoidea	Ordovician,	Permian, Lopingian
	Crinoidea	Cambrian, Middle	Holocene
	Asteroidea	Ordovician, Early	Holocene
	Holothuroidea	Ordovician, Early	Holocene
	Echinoidea	Ordovician, Late	Holocene
Graptolithina		Cambrian, Furongian	Carboniferous, Late Mississippian
	Dendroidea	Cambrian, Furongian	Carboniferous, Late Mississippian
	Graptoloidea	Ordovician, Early	Devonian, Early
Chordata	*Conodonta*	Cambrian, Late	Triassic, Late
	Reptilia	Carboniferous, Pennsylvanian	Holocene
	Testudines	Triassic, Late	Holocene
	Ichthyosauria	Triassic, Early	Cretaceous, Late
	Squamata	Jurassic, Late	Holocene

7

Table 7.10 (*cont.*)

Kingdom or Phylum[a]	Class[a]	Subclass, superorder or order[a]	Range First	Last
	Reptilia (*cont.*)	Crocodylia	Triassic, Late	Holocene
		Pterosauria	Triassic, Late	Cretaceous, Late
		Saurischia	Triassic, Late	Cretaceous, Late
		Ornithischia	Triassic, Late	Cretaceous, Late
		Sauropterygia	Triassic, Early	Cretaceous, Late
		Synapsida: Pelycosauria	Carboniferous, Pennsylvanian	Permian, Guadalupian
		Synapsida: Therapsida	Permian, Cisuralian	Jurassic, Middle
	Aves		Jurassic, Late	Holocene
		Archaeopterygiformes	Jurassic, Late	Cretaceous, Early
	Mammalia		Triassic, Late	Holocene
		Prototheria	Cretaceous, Early	Holocene
		Triconodonta	Triassic, Late	Cretaceous, Late
		Allotheria	Triassic, Late	Eocene, Late
		Marsupialia	Cretaceous, Late	Holocene
		Placentalia, Edentata	Paleocene, Late	Holocene
		Placentalia, Epitheria	Cretaceous, Early	Holocene

7

Group		
Rodentia	Paleocene, Late	Holocene
Creodonta	Paleocene, Late	Miocene, Late
Carnivora	Paleocene, Late	Holocene
Lipotyphla	Paleocene, Early	Holocene
Chiroptera	Paleocene, Late	Holocene
Primates	Paleocene, Early	Holocene
Artiodactyla	Eocene, Early	Holocene
Cetacea	Eocene, Early	Holocene
Notoungulata	Paleocene, Late	Pleistocene
Condylarthra	Cretaceous, Late	Oligocene, Early
Proboscidea	Paleocene, Late	Holocene
Perissodactyla	Paleocene, Early	Holocene

Explanation and notes: This table gives the stratigraphical ranges of selected fossil groups and is entirely based on the data presented in *The Fossil Record 2* (1993). No attempt has been made to adapt the taxonomic terms or relationships, even though they are sometimes obsolete, partly for reasons of practicality and partly so the table can be used more readily in conjunction with the original work, if desired. Stratigraphic divisional names have been adapted to bring them into line with current usage. The Preface to *The Fossil Record 2* comments on the potential pitfalls of synthesising data of this kind. This table must also be used in the knowledge that it could be subject to criticism by different scholars but in the knowledge that the original source was, and still probably is, the best of its sort available. This table is just a selection – again based on practicalities of compilation but also judged to be of use to students and others. *The Fossil Record 2* includes much more information, including groups (e.g. Bryozoa) not in this list. See also Table 7.9, which includes data on other groups including fish and plants. *a* In a few cases a suitable equivalent is given to the phylum, class, subclass or order – see the source for precise groupings.

Source: Benton, M.J. ed. (1993). *The Fossil Record 2*. London: Chapman & Hall.

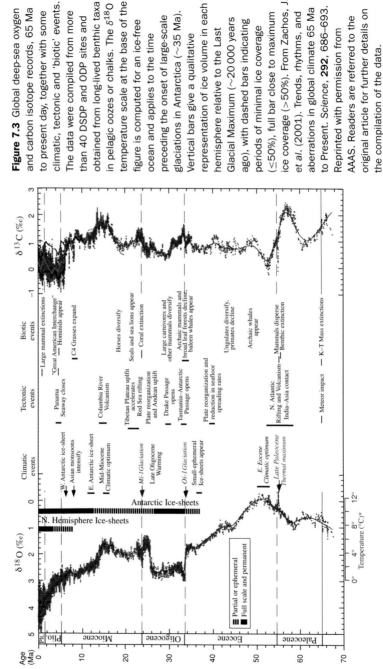

Figure 7.3 Global deep-sea oxygen and carbon isotope records, 65 Ma to present day, together with some climatic, tectonic and 'biotic' events. The data were compiled from more than 40 DSDP and ODP sites and obtained from long-lived benthic taxa in pelagic oozes or chalks. The $\delta^{18}O$ temperature scale at the base of the figure is computed for an ice-free ocean and applies to the time preceding the onset of large-scale glaciations in Antarctica (~35 Ma). Vertical bars give a qualitative representation of ice volume in each hemisphere relative to the Last Glacial Maximum (~20 000 years ago), with dashed bars indicating periods of minimal ice coverage (≤50%), full bar close to maximum ice coverage (>50%). From Zachos, J. et al. (2001). Trends, rhythms, and aberrations in global climate 65 Ma to Present. *Science*, **292**, 686–693. Reprinted with permission from AAAS. Readers are referred to the original article for further details on the compilation of the data.

174

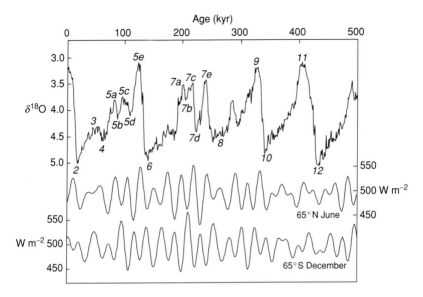

Figure 7.4 Orbital timescale climate change. The record of marine benthic carbonate oxygen isotopes for the past 500 thousand years, with curves of incoming solar radiation (insolation) for 65°N June and 65°S December. $\delta^{18}O$ (versus the PDB standard) from: Lisiecki, L. E. and Raymo, M. E. (2005). A Pliocene–Pleistocene stack of 57 globally distributed benthic $\delta^{18}O$ records. *Palaeoceanography,* **20**, doi: 10.1029/2004PA001071. Refer to that reference for the full isotope stack covering the last 5 Ma. Insolation from: Laskar, J. *et al.* (2004). A long-term numerical solution for the insolation quantities of the Earth. *Astronomy & Astrophysics,* **428**, 261–285. Numbers on the $\delta^{18}O$ curve refer to marine isotope stages and substages (MIS). MIS 2 is equivalent to the last glacial period, MIS5e to the last interglacial. MIS refers to a period of time between climate boundaries. Maxima and minima on the curve are also sometimes referred to numerically (e.g. 5.5 is the peak of the last interglacial).

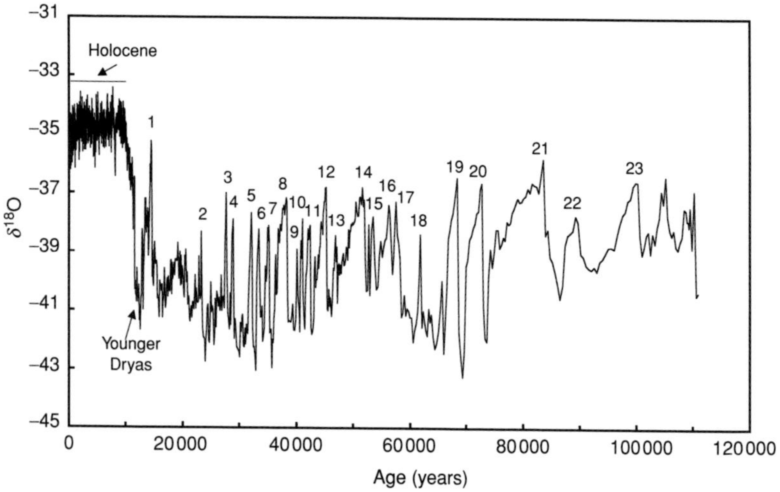

Figure 7.5 Millennial scale climate variability of the last 100 thousand years as recorded by oxygen isotopes of ice in central Greenland. $\delta^{18}O$ (versus the SMOW standard) from the GISP2 ice core (Stuiver, M. and Grootes, P. M. (2000). GISP2 oxygen isotope ratios. *Quaternary Research*, **53**, 277–284). The Holocene and Younger Dryas periods are marked, along with numbered interstadials following Dansgaard, W. *et al.* (1993). Evidence for general instability of past climate from a 250-kyr ice-core record. *Nature*, **364**, 218–220.

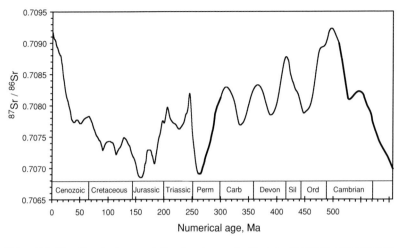

Figure 7.6 Strontium isotope evolution line (oceans) with geological time. The variation in the strontium isotope ratio of strontium dissolved in the World's oceans can be used to correlate and date some sediments using marine minerals including certain biogenic ones. It is also possible to measure the duration of stratigraphic gaps and the duration of biozones etc. Updated by J. M. McArthur (2008) from McArthur, J. M. and Howarth, R. J. (2004). Strontium isotope stratigraphy. In *A Geologic Time Scale 2004*, ed. F. M. Gradstein *et al.* Cambridge: Cambridge University Press, with permission of the authors.

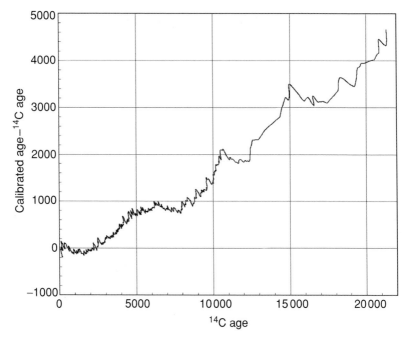

Figure 7.7 Calibration curve for radiocarbon ages, following IntCal04. Differences in the calibrated (or calendar) age from the raw ¹⁴C age are plotted against raw ¹⁴C ages. Note that through most of this time span ¹⁴C ages are too young. All ages are relative to 1950, following normal ¹⁴C convention in which 1950 is defined as 'the present'. Automatic calibration of ¹⁴C ages and other details of radiocarbon chronology can be found at http://c14.arch.ox.ac.uk/

Source: Reimer, P. J. *et al.* (2004). IntCal04 terrestrial radiocarbon age calibration, 0–26 cal kyr BP. *Radiocarbon*, **46**, 1029–1058.

8 Chemistry and isotopes

8

Table 8.1 The periodic table with atomic weights

1	2	3	4	5	6	7	8	9	10	11	12	13	14	15	16	17	18
1 **H** 1.007 94																	2 **He** 4.002 602
3 **Li** 6.941	4 **Be** 9.012 182											5 **B** 10.811	6 **C** 12.0107	7 **N** 14.0067	8 **O** 15.9994	9 **F** 18.998 4032	10 **Ne** 20.1797
11 **Na** 22.989 770	12 **Mg** 24.305											13 **Al** 26.981 538	14 **Si** 28.0855	15 **P** 30.973 761	16 **S** 32.065	17 **Cl** 35.453	18 **Ar** 39.948
19 **K** 39.0983	20 **Ca** 40.078	21 **Sc** 44.955 910	22 **Ti** 47.867	23 **V** 50.9415	24 **Cr** 51.9961	25 **Mn** 54.938 049	26 **Fe** 55.845	27 **Co** 58.933 200	28 **Ni** 58.6934	29 **Cu** 63.546	30 **Zn** 65.409	31 **Ga** 69.723	32 **Ge** 72.64	33 **As** 74.921	34 **Se** 78.96	35 **Br** 79.904	36 **Kr** 83.798
37 **Rb** 85.4678	38 **Sr** 87.62	39 **Y** 88.905 85	40 **Zr** 91.224	41 **Nb** 92.906 38	42 **Mo** 95.94	43 **Tc** 97.9072	44 **Ru** 101.07	45 **Rh** 102.905 50	46 **Pd** 106.42	47 **Ag** 107.8682	48 **Cd** 112.411	49 **In** 114.818	50 **Sn** 118.710	51 **Sb** 121.760	52 **Te** 127.60	53 **I** 126.904 47	54 **Xe** 131.293
55 **Cs** 132.905 45	56 **Ba** 137.327	57–71 lanthanoids	72 **Hf** 178.49	73 **Ta** 180.9479	74 **W** 183.84	75 **Re** 186.207	76 **Os** 190.23	77 **Ir** 192.217	78 **Pt** 195.078	79 **Au** 196.966 55	80 **Hg** 200.59	81 **Tl** 204.3833	82 **Pb** 207.2	83 **Bi** 208.890 38	84 **Po** (208.9824)	85 **At** (209.9871)	86 **Rn** (222.0176)
87 **Fr** (223.0197)	88 **Ra** (226.0254)	89–103 actinoids	104 **Rf** (261.1088)	105 **Db** (262.1141)	106 **Sg** (266.1219)	107 **Bh** (264.12)	108 **Hs** (277)	109 **Mt** (268.1388)	110 **Ds** (271)	111 **Rg** (272)							

57 **La** 138.9055	58 **Ce** 140.116	59 **Pr** 140.907 65	60 **Nd** 144.24	61 **Pm** (144.9127)	62 **Sm** 150.36	63 **Eu** 151.964	64 **Gd** 157.25	65 **Tb** 158.925 34	66 **Dy** 162.500	67 **Ho** 164.930 32	68 **Er** 167.259	69 **Tm** 168.934 21	70 **Yb** 173.04	71 **Lu** 174.967
89 **Ac** (227.0277)	90 **Th** 232.0381	91 **Pa** 231.035 88	92 **U** 238.028 91	93 **Np** (237.0482)	94 **Pu** (244.0642)	95 **Am** (243.0614)	96 **Cm** (247.0704)	97 **Bk** (247.0703)	98 **Cf** (251.0796)	99 **Es** (252.0830)	100 **Fm** (257.0951)	101 **Md** (258.0984)	102 **No** (259.1010)	103 **Lr** (262.1097)

Sources: International Union of Pure and Applied Chemistry, 1 November 2004.

8

Table 8.2 Ionic radii for given co-ordination numbers and spin state[a]

Ion	CN and spin	Radius (10^{-1} nm)	Ion	CN and spin	Radius (10^{-1} nm)
Ag^+	II	0.67	Ca^{2+}	VII	1.06
	IV sq	1.02		VIII	1.12
	V	1.09		IX	1.18
	VI	1.15		X	1.23
	VII	1.22		XII	1.34
	VIII	1.28	Cd^{2+}	IV	0.78
Al^{3+}	IV	0.39		V	0.87
	V	0.48		VI	0.95
	VI	0.535		VII	1.03
As^{5+}	IV	0.335		VIII	1.10
	VI	0.46		XII	1.31
Au^{3+}	IV sq	0.68	Ce^{3+}	VI	1.01
B^{3+}	III	0.01		VII	1.07
	IV	0.11		VIII	1.143
	VI	0.27		IX	1.196
Ba^{2+}	VI	1.35		X	1.25
	VII	1.38		XII	1.34
	VIII	1.42	Ce^{4+}	VI	0.87
	IX	1.47		VIII	0.97
	X	1.52		X	1.07
	XI	1.57		XII	1.14
	XII	1.61	Cl^{7+}	IV	0.08
Be^{2+}	III	0.16	Co^{2+}	IVH	0.58
	IV	0.27		V	0.67
	VI	0.45		VIL	0.65
Bi^{3+}	V	0.96		VIH	0.745
	VI	1.03		VIII	0.90
	VIII	1.17	Cr^{z+}	VIL	0.73
Br^-	VI	1.96		VIH	0.80
Br^{7+}	IV	0.25	Cr^{3+}	VI	0.615
Ca^{2+}	VI	1.00	Cs^+	VI	1.67

(*cont.*)

Table 8.2 (*cont.*)

Ion	CN and spin	Radius (10^{-1} nm)	Ion	CN and spin	Radius (10^{-1} nm)
Cs^+	VIII	1.74	F^-	IV	1.31
	IX	1.78		VI	1.33
	X	1.81	Fe^{2+}	IVH	0.63
	XI	1.85		IV sq H	0.64
	XII	1.88		VIL	0.61
Cu^+	II	0.46		VIH	0.780
	IV	0.60		VIII	0.92
	VI	0.77	Fe^{3+}	IVH	0.49
Cu^{2+}	IV sq	0.57		V	0.58
	IV	0.57		VIL	0.55
	V	0.65		VIH	0.645
	VI	0.73		VIIIH	0.78
Dy^{3+}	VI	0.912	Ga^{3+}	IV	0.47
	VII	0.97		V	0.55
	VIII	1.027		VI	0.620
	IX	1.083	Gd^{3+}	VI	0.938
Er^{3+}	VI	0.890		VII	1.00
	VII	0.945		VIII	1.053
	VIII	1.004		IX	1.107
	IX	1.062	Ge^{4+}	IV	0.390
Eu^{2+}	VI	1.17		VI	0.530
	VII	1.20	Hf^+	IV	0.58
	VIII	1.25		VI	0.71
	IX	1.30		VII	0.76
	X	1.35		VIII	0.83
Eu^{3+}	VI	0.947	Hg^+	III	0.97
	VII	1.01		VI	1.19
	VIII	1.066	Hg^{2+}	II	0.69
	IX	1.120		IV	0.96
F^-	II	1.285		VI	1.02
	III	1.30		VIII	1.14

Table 8.2 (*cont.*)

Ion	CN and spin	Radius (10^{-1} nm)	Ion	CN and spin	Radius (10^{-1} nm)
Ho^{3+}	VI	0.901	Mg^{2+}	VIII	0.89
	VIII	1.015	Mn^{2+}	IV*H*	0.66
	IX	1.072		V*H*	0.75
	X	1.12		VI*L*	0.67
In^{3+}	IV	0.62		VI*H*	0.830
	VI	0.800		VII*H*	0.90
	VIII	0.92		VIII	0.96
Ir^{3+}	VI	0.68	Mn^{4+}	IV	0.39
Ir^{4+}	VI	0.625		VI	0.530
K^+	IV	1.37	Mo^{4+}	VI	0.650
	VI	1.38	Mo^{6+}	IV	0.41
	VII	1.46		V	0.50
	VIII	1.51		VI	0.59
	IX	1.55		VII	0.73
	X	1.59	Na^+	IV	0.99
	XII	1.64		V	1.00
La^{3+}	VI	1.032		VI	1.02
	VII	1.10		VII	1.12
	VIII	1.160		VIII	1.18
	IX	1.216		IX	1.24
	X	1.27		XII	1.39
	XII	1.36	Nb^{5+}	IV	0.48
Li^+	IV	0.590		VI	0.64
	VI	0.76		VII	0.69
	VIII	0.92		VIII	0.74
Lu^{3+}	VI	0.861	Nd^{3+}	VI	0.983
	VIII	0.977		VIII	1.109
	IX	1.032		IX	1.163
Mg^{2+}	IV	0.57		XII	1.27
	V	0.66	Ni^{2+}	IV	0.55
	VI	0.720		IV sq	0.49

(*cont.*)

Table 8.2 (*cont.*)

Ion	CN and spin	Radius $(10^{-1}$ nm$)$	Ion	CN and spin	Radius $(10^{-1}$ nm$)$
Ni^{2+}	V	0.63	Pd^{4+}	VI	0.615
	VI	0.690	Pr^{3+}	VI	0.99
O^{2-}	II	1.35		VIII	1.126
	III	1.36		IX	1.179
	IV	1.38	Pt^{2+}	IV sq	0.60
	VI	1.40	Pt^{4+}	VI	0.625
	VIII	1.42	Ra^{2+}	VIII	1.48
Os^{4+}	VI	0.630		XII	1.70
P^{5+}	IV	0.17	Rb^+	VI	1.52
	V	0.29		VII	1.56
	VI	0.38		VIII	1.61
Pa^{4+}	VI	0.90		IX	1.63
	VIII	1.01		X	1.66
Pa^{5+}	VI	0.78		XI	1.69
	VIII	0.91		XII	1.72
	IX	0.95		XIV	1.83
Pb^{2+}	IV py	0.98	Re^{4+}	VI	0.63
	VI	1.19	Re^{5+}	VI	0.58
	VII	1.23	Re^{6+}	VI	0.55
	VIII	1.29	Rh^{3+}	VI	0.665
	IX	1.35	Rh^{4+}	VI	0.60
	X	1.40	Ru^{3+}	VI	0.68
	XI	1.45	Ru^{4+}	VI	0.620
	XII	1.49	S^{6+}	IV	0.12
Pb^{4+}	IV	0.65		VI	0.29
	V	0.73	Sb^{3+}	IV py	0.76
	VI	0.775		V	0.80
	VIII	0.94		VI	0.76
Pd^+	II	0.59	Sb^{5+}	VI	0.60
Pd^{2+}	IV sq	0.64	Sc^{3+}	VI	0.745
	VI	0.86		VIII	0.870

Table 8.2 (*cont.*)

Ion	CN and spin	Radius (10⁻¹ nm)	Ion	CN and spin	Radius (10⁻¹ nm)
Se^{6+}	IV	0.28	Th^{4+}	VIII	1.05
	VI	0.42		IX	1.09
Si^{4+}	IV	0.26		X	1.13
	VI	0.400		XI	1.18
Sm^{3+}	VI	0.958		XII	1.21
	VII	1.02	Ti^{3+}	VI	0.670
	VIII	1.079	Ti^{4+}	IV	0.42
	IX	1.132		V	0.51
	XII	1.24		VI	0.605
Sn^{4+}	IV	0.55		VIII	0.74
	V	0.62	Tl^{+}	VI	1.50
	VI	0.690		VIII	1.59
	VII	0.75		XII	1.70
	VIII	0.81	Tl^{3+}	IV	0.75
Sr^{2+}	VI	1.18		VI	0.885
	VII	1.21		VIII	0.98
	VIII	1.26	Tm^{3+}	VI	0.880
	IX	1.31		VIII	0.994
	X	1.36		IX	1.052
	XII	1.44	U^{4+}	VI	0.89
Ta^{5+}	VI	0.64		VII	0.95
	VII	0.69		VIII	1.00
	VIII	0.74		IX	1.05
Tb^{3+}	VI	0.923		XII	1.17
	VII	0.98	U^{5+}	VI	0.76
	VIII	1.040		VII	0.84
	IX	1.095	U^{6+}	II	0.45
Te^{4+}	III	0.52		IV	0.52
	IV	0.66	U^{6+}	VI	0.73
	VI	0.97		VII	0.81
Th^{4+}	VI	0.94		VIII	0.86

(*cont.*)

Table 8.2 (*cont.*)

Ion	CN and spin	Radius $(10^{-1}$ nm)	Ion	CN and spin	Radius $(10^{-1}$ nm)
V^{3+}	VI	0.640	Zn^{2+}	IV	0.60
W^{4+}	VI	0.66		V	0.68
W^{6+}	IV	0.42		VI	0.740
	V	0.51		VIII	0.90
	VI	0.60	Zr^{4+}	IV	0.59
Y^{3+}	VI	0.900		V	0.66
	VII	0.96		VI	0.72
	VIII	1.019		VII	0.78
	IX	1.075		VIII	0.84
Yb^{3+}	VI	0.868		IX	0.89
	VII	0.925			
	VIII	0.985			
	IX	1.042			

Notes: *a* Data are given for those oxidation states and co-ordination numbers (CN) that are likely to be relevant to natural Earth systems. py = square pyramidal, sq = square planar, H = high spin, L = low spin state.

Sources: Henderson, P. (1982). *Inorganic Geochemistry.* Oxford: Pergamon Press. Shannon, R. D. (1976). *Acta Crystallographica A*, **32**, 751–765.

Table 8.3 Ionisation enthalpies of selected elements[a]

		First	Second	Third	Fourth
1	H	1311			
3	Li	520.0			
4	Be	899.1	1758		
5	B	800.5	2428	2394	25 020
6	C	1086	2353	4618	6 512
11	Na	495.8			
12	Mg	737.5	1450		
13	Al	577.5	1817	2745	
14	Si	786.3	1577	3228	
15	P	1012	1903	2910	4 955
16	S	999.3	2260	3380	4 562
19	K	418.7			
20	Ca	589.6	1146		
21	Sc	631	1235	2389	
22	Ti	656	1309	2650	4 173
23	V	650	1414	2828	4 600
24	Cr	652.5	1592	3056	4 900
25	Mn	717.1	1509	3251	
26	Fe	762	1561	2956	
27	Co	758	1644	3231	
28	Ni	736.5	1752	3489	
29	Cu	745.2	1958	3545	
30	Zn	906.1	1734	3831	
31	Ga	579	1979	2962	
32	Ge	760	1537	3301	4 410
33	As	947	1798	2735	4 830
34	Se	941	2070	3090	4 140
37	Rb	402.9	2650	3900	
38	Sr	549.3	1064		
39	Y	616	1180	1979	
40	Zr	674.1	1268	2217	3 313
41	Nb	664	1381	2416	3 700
42	Mo	685	1558	2618	4 480

(*cont.*)

Table 8.3 (*cont.*)

		First	Second	Third	Fourth
43	Tc	703	1472	2850	
44	Ru	710.6	1617	2746	
45	Rh	720	1744	2996	
46	Pd	804	1874	3177	
47	Ag	730.8			
48	Cd	876.4	1630		
49	In	558.1	1820	2705	
50	Sn	708.2	1411	2942	3 928
51	Sb	833.5	1590	2440	4 250
52	Te	869	1800	3000	3 600
55	Cs	375.5			
56	Ba	502.5	964		
57	La	541	1103	1849	
72	Hf	760	1440	2250	3 210
73	Ta	760	1560		
74	W	770	1710		
78	Pt	870	1791		
79	Au	889	1980		
80	Hg	1007	1809	3300	
81	Tl	588.9	1970	2880	4 890
82	Pb	715.3	1450	3080	4 082
83	Bi	702.9	1609	2465	4 370

Note: *a* Units kJ mol^{-1}. Ionisation enthalpies are now more commonly used than 'ionisation potentials', which are the negative of the enthalpy and expressed in units of electron volts.

Source: Cotton, F. A. *et al*. (1999). *Advanced Inorganic Chemistry*, 6th edn. New York: John Wiley.

Table 8.4 Electronegativities of the elements

Z	Element	EN[a]	Z	Element	EN[a]
3	Li	0.98	32	Ge	2.01
4	Be	1.57	33	As	2.18
5	B	2.04	34	Se	2.55
6	C	2.55	35	Br	2.96
7	N	3.04	37	Rb	0.82
8	O	3.44	38	Sr	0.95
9	F	3.98	39	Y	1.22
11	Na	0.93	40	Zr	1.33
12	Mg	1.31	42	Mo	2.16
13	Al	1.61	45	Rh	2.28
14	Si	1.90	46	Pd	2.20
15	P	2.19	47	Ag	1.93
16	S	2.58	48	Cd	1.69
17	Cl	3.16	49	In	1.78
19	K	0.82	50	Sn	1.96
20	Ca	1.00	51	Sb	2.05
21	Sc	1.36	53	I	2.66
22	Ti	1.54	55	Cs	0.79
23	V	1.63	56	Ba	0.89
24	Cr	1.66	74	W	2.36
25	Mn	1.55	77	Ir	2.20
26	Fe	1.83	78	Pt	2.28
27	Co	1.88	79	Au	2.54
28	Ni	1.91	80	Hg	2.00
29	Cu	1.90	81	Tl	2.04
30	Zn	1.65	82	Pb	2.33
31	Ga	1.81	83	Bi	2.02

Note: a The values given here are Pauling values. Estimates by other authors exist but the Pauling ones are often quoted. Pauling describes electronegativity as the power of an atom in a molecule to attract electrons to itself. The units approximate to electron volts but the figures are estimates and are modified to provide consistency. The values given are for the following common oxidation states: columns 1 and 17 of the periodic table (see Table 8.1), oxidation state of 1; columns 3, 13 and 15, state 3; columns 4 and 14 state 4; the rest, state 2.

For discussion on electronegativity see Atkins, P., *et al.* (2006). *Shriver – Atkins Inorganic Chemistry*, 4th edn. Oxford: Oxford University Press; Henderson, P. (1982). *Inorganic Geochemistry*. Oxford: Pergamon Press.

Sources: as above plus: Pauling, L. (1960). *The Nature of the Chemical Bond,* 3rd edn. Ithaca, NY: Cornell University Press.

Table 8.5 Valency states: elements in terrestrial minerals

Element	1	2	3	4	5	6	Element	1	2	3	4	5	6
Li	+						Rb	+					
Be		+					Sr		+				
B			+				Y			+			
C				+			Zr				+		
O		−					Nb					+	
F	−						Mo						+
Na	+						Ru		+		+		
Mg		+					Rh			+			
Al			+				Pd		+		+		
Si				+			Ag	+					
P					+		Cd		+				
S		−					In			+			
Cl	−						Sn		+		+		
K	+						Sb			+		(+)	
Ca		+					Te	−			+		
Sc			+				I	−					
Ti			(+)	+			Cs	+					
V			+				Ba		+				
Cr			+				La			+			
Mn		+					Ce			+	(+)		
Fe		+	+				Pr			+			
Co		+					Nd			+			
Ni		+					Sm			+			
Cu	+	+					Eu		+	+			
Zn		+					Gd			+			
Ga			+				Tb			+			
Ge				+			Dy			+			
As			+		+		Ho			+			
Se	−			+			Er			+			
Br	−						Tm			+			

Table 8.5 (*cont.*)

Element	\- Valency state 1	2	3	4	5	6	Element	\- Valency state 1	2	3	4	5	6
Yb			+				Pt		+		+		
Lu			+				Au			+			
Hf				+			Hg	+	+				
Ta					+		Tl	+		+			
W						+	Pb		+				
Re				+			Bi			+			
Os				+			Th				+		
Ir			+				U				+	+	+

Note: Some elements have other oxidation states in non-terrestrial minerals e.g. Cr^{2+} may exist in lunar minerals. The values are not necessarily applicable to aqueous and sedimentary systems. The oxidation states are formal in the cases of those elements that are covalently bonded in many minerals. Some elements also occur in the 'native' state. Parentheses indicate rare oxidation states in unusual redox conditions.

Source: modified from Henderson, P. (1982). *Inorganic Geochemistry.* Oxford: Pergamon Press, with permission.

8

Table 8.6a Crystal Field Stabilisation Energies (CFSEs)

Ion	Electronic config.[a]	3d electron configuration					CFSE

Octahedral co-ordination

Ion	Electronic config.[a]	t_{2g}			e_g		CFSE
Sc^{3+}; Ti^{4+}	(Ar)						0
Ti^{3+}	$(Ar)3d^1$	↑					$2/5\Delta_0$
V^{3+}	$(Ar)3d^2$	↑	↑				$4/5\Delta_0$
Cr^{3+}	$(Ar)3d^3$	↑	↑	↑			$6/5\Delta_0$
Mn^{2+}; Fe^{3+}	$(Ar)3d^5$	↑	↑	↑	↑	↑	0
Fe^{2+}	$(Ar)3d^6$	↑↓	↑	↑	↑	↑	$2/5\Delta_0$
Co^{2+}	$(Ar)3d^7$	↑↓	↑↓	↑	↑	↑	$4/5\Delta_0$
Ni^{2+}	$(Ar)3d^8$	↑↓	↑↓	↑↓	↑	↑	$6/5\Delta_0$
Cu^{2+}	$(Ar)3d^9$	↑↓	↑↓	↑↓	↑↓	↑	$3/5\Delta_0$
Zn^{2+}	$(Ar)3d^{10}$	↑↓	↑↓	↑↓	↑↓	↑↓	0

Tetrahedral co-ordination

Ion	Electronic config.[a]	e		t_2			CFSE
Sc^{3+}; Ti^{4+}	(Ar)						
Ti^{3+}	$(Ar)3d^1$	↑					$3/5\Delta_t$
V^{3+}	$(Ar)3d^2$	↑	↑				$6/5\Delta_t$
Cr^{3+}	$(Ar)3d^3$	↑	↑	↑			$4/5\Delta_t$
Mn^{2+}; Fe^{3+}	$(Ar)3d^5$	↑	↑	↑	↑	↑	0
Fe^{2+}	$(Ar)3d^6$	↑↓	↑	↑	↑	↑	$3/5\Delta_t$
Co^{2+}	$(Ar)3d^7$	↑↓	↑↓	↑	↑	↑	$6/5\Delta_t$
Ni^{2+}	$(Ar)3d^8$	↑↓	↑↓	↑↓	↑	↑	$4/5\Delta_t$
Cu^{2+}	$(Ar)3d^9$	↑↓	↑↓	↑↓	↑↓	↑	$2/5\Delta_t$
Zn^{2+}	$(Ar)3d^{10}$	↑↓	↑↓	↑↓	↑↓	↑↓	0

Notes: a Electronic config. = electronic configuration; (Ar) = configuration of argon: $1s^2 2s^2 2p^6 3s^2 3p^6$. The above CFSEs are for the high-spin state; the low-spin state is rare or absent in minerals in the Earth's crust.

Sources: Burns, R. G. (1993). *Mineralogical Applications of Crystal Field Theory*, 2nd edn. Cambridge: Cambridge University Press. Henderson, P. (1982). *Inorganic Geochemistry*. Oxford: Pergamon Press.

Table 8.6b Crystal Field Stabilisation Energy (CFSE) and Octahedral Site Preference Energy (OSPE) for transition metal ions in spinels

No. of 3d electrons	Ion	Octahedral CFSE (kJ mol^{-1})	Tetrahedral CFSE (kJ mol^{-1})	OSPE (kJ mol^{-1})
0	Sc^{3+}, Ti^{4+}	0	0	0
1	Ti^{3+}	87.5	58.6	28.9
2	V^{3+}	160.2	106.7	53.5
3	Cr^{3+}	224.7	66.9	157.8
4	Cr^{2+}	100.4	29.3	71.1
5	Mn^{2+}, Fe^{3+}	0	0	0
6	Fe^{2+}	49.8	33.1	16.7
7	Co^{2+}	92.9	61.9	31.0
8	Ni^{2+}	122.2	36.0	86.2
9	Cu^{2+}	90.4	26.8	63.6
10	Zn^{2+}	0	0	0

Note: The octahedral site preference energy is the difference between the octahedral CFSE and tetrahedral CFSE of a given ion. This table gives data for spinels – values will be different for different minerals but the data are a useful guideline.

Sources: Dunitz, J. D. and Orgel, L. E. (1957). Electronic properties of transition-metal oxides. II Cation distribution amongst octahedral and tetrahedral sites. *Journal of Physics and Chemistry of Solids*, **3**, 318–323. McClure, D. S. (1957). The distribution of transition metal cations in spinels. *Journal of Physics and Chemistry of Solids*, **3**, 311–317.

Table 8.7a Mineral/melt distribution coefficients for basaltic and andesitic rocks[a]

Z	Olivine	Ortho-pyroxene	Clino-pyroxene	Amphibole	Plagioclase	Phlogopite	Garnet
2 He	0.1						
10 Ne	0.08						
11 Na	0.02	0.08	0.23	0.7	1.2		
12 Mg	9	6	3.5	7	0.04		
13 Al		0.3	0.4	0.8	1.9		
18 Ar	0.03						
19 K	0.007[b]	0.01	0.03	0.60	0.17	3	
20 Ca	0.03	0.3					
21 Sc	0.17	1.2	2.9	3.2	0.03		4
22 Ti	0.03	0.1	0.8		0.04	0.90	0.4
23 V	0.06	0.6	1.3	3.4	0.01		
24 Cr	0.7	10	9	2[b]	0.08		
25 Mn	1.2	1.4	0.9		0.05		
26 Fe^{2+}	1.9	1.8	0.8				
27 Co	4.7	3	1.2	1.7	0.06		1.1
28 Ni	18	5	2.6	6.8			
31 Ga			0.4		1.0		
36 Kr	0.1[b]						
37 Rb	0.006[b]	0.02	0.04[b]	0.25[b]	0.10[b]	3	
38 Sr	0.01	0.01	0.14	0.6	1.8	0.1	0.01
54 Xe	0.2[b]						
55 Cs	0.0004		0.01	0.1	0.03		

8

56 Ba	0.006^b	0.013	0.07^b	0.3	0.23^b	1.1	0.06
57 La	0.007		0.13	0.3	0.13		0.04
58 Ce	0.008	0.02^b	0.2	0.3	0.14	0.03	
60 Nd	0.007	0.04	0.3	0.3	0.08^b	0.03	
62 Sm	0.009	0.05	0.5	0.9	0.08^b	0.03	0.6
63 Eu	0.01	0.05	0.8	1.1	0.3	0.03	0.9
64 Gd	0.01	0.09	0.7	1.1	0.1		
66 Dy	0.01	0.15	1.1		0.09^b	0.03	
68 Er	0.01	0.2	1.0		0.08	0.03	5.5
70 Yb	0.01	0.34	1.0	1.0	0.07		30
71 Lu	0.03	0.2	0.7	0.9	0.08	0.04	35
72 Hf	0.01		0.3	0.9	0.05		0.4
82 Pb					0.26		
90 Th	0.01		0.05^b		0.05		
92 U	0.003		0.04^b	0.1	0.01		

Notes: *a* The data are guidance values only and are based on the concentration ratio, by weight, of the element in the mineral to that in the matrix of the igneous rock. The matrix composition is considered to represent that of the magma from which the minerals crystallised. The coefficients are, therefore, for real case situations and are not necessarily those prevailing at thermodynamic equilibrium. The tables do not include experimentally acquired data. *b* indicates that the range of values used to obtain the average or 'guidance' figure is over one or more orders of magnitude.

Sources: Mainly: Lemarchand, F. *et al.* (1987). Trace element distribution coefficients in alkaline series. *Geochimica et Cosmochimica Acta*, **51**, 1071–1081; Henderson, P. (1982). *Inorganic Geochemistry*. Oxford: Pergamon Press; Rollinson, H. (1993). *Using Geochemical Data: Evaluation, Presentation, Interpretation*. Harlow: Longman; Valbracht, P. J. *et al.* (1994). *Noble Gas Geochemistry*, ed. J. Matsuda. Tokyo: Terra Scientific Publishing Co.; Villement, B. *et al.* (1981). Distribution coefficients of major and trace elements; fractional crystallization in the alkali basalt series of Chaine des Puys (Massif Central, France). *Geochimica et Cosmochimica Acta*, **45**, 1997–2016.

8

Table 8.7b Mineral/melt distribution coefficients for dacitic and rhyolitic rocks

Z	Ortho-pyroxene	Clino-pyroxene	Amphibole	Plagioclase	Alkali feldspar	Biotite	Garnet	Magnetite	Apatite	Zircon
11 Na	0.06	0.1		1.5						
12 Mg				0.3	0.2					
19 K	0.3[b]	0.04	0.08	0.2		4	0.1			
20 Ca	1.4	10	1.8		2[b]	0.6		0.4		
21 Sc	14	45	20	0.07[b]	0.03	13	16	4		
22 Ti	0.4	0.7	7	0.05			1.2		0.1	
23 V	6									
24 Cr	1.6	30		0.2[b]	0.2[b]	12		11		
25 Mn	40	8		0.2[b]						
27 Co	3			0.08	0.2		2.6	30		
30 Zn	0.9			0.4				12		
37 Rb	0.09[b]	0.06	0.01	0.09[b]	0.4	3.3	0.01			
38 Sr	0.02	0.5	0.02	6[b]	9	0.26	0.02			
39 Y	1	3.5	6	0.1					40	
40 Zr	0.1	0.1		0.1	0.03	1	1.2	4	0.4	
41 Nb	0.8	0.8	4	0.06		6			0.1	
55 Cs				0.08	0.1	2.5				
56 Ba	0.02[b]	0.1[b]	0.04	0.5	6	10	0.02	0.07		

8

57 La	0.7	0.6	0.8	0.35	0.09	3.2[b]	0.39	0.5	14	4.2
58 Ce	0.35[b]	0.7[b]	1.1	0.24	0.05	1.4[b]	0.4	0.6	31	4.2
60 Nd	0.6[b]	2[b]	2.8	0.2	2.6[b]	0.9[b]	0.6	0.9	50	4.3
62 Sm	0.6[b]	2[b]	5	0.1[b]	0.02	0.7[b]	2.2	0.9	54	4.7
63 Eu	0.4	1.7[b]	3.6	2.7	2.7	0.5[b]	0.7[b]	0.6	27	3.4
64 Gd	0.5[b]	2[b]	6[b]	0.26	0.01	0.2	7.5[b]		40	6.6
65 Tb	1.4	4.5		0.15		0.03	12	0.8		
66 Dy	0.6[b]	3	8	0.1[b]	0.04	0.5[b]		0.8	42	48
68 Er	0.5	1	7	0.06	0.01	0.25	43		31	140
69 Tm	1.4		5	0.1				0.8		
70 Yb	0.9	2.2	4[b]	0.09[b]	0.02	0.5	38	0.4	21	280
71 Lu	1.0	1.4	4	0.06	0.02	0.6	34	0.4	17	340
72 Hf	0.1	0.4		0.07	0.02	1.1	3.3	0.3	0.7	
73 Ta	0.5	0.4	0.7	0.05	0.01	1.4		1.5		48
82 Pb				0.5	1.4	0.8				
90 Th	0.15	0.13[b]	0.2	0.04	0.02	0.9		0.1		77
92 U	0.1	0.03	0.4	0.03	0.03	0.5		0.2		340

Notes and sources: See Table 8.7a.

Table 8.8 Thermodynamic properties of selected minerals

Mineral	Free energy of formation $\Delta_f G$ (kJ mol^{-1})	Enthalpy of formation $\Delta_f H$ (kJ mol^{-1})	Entropy S (J mol^{-1} K^{-1})
Elements			
Carbon	0	0	5.74
Diamond	2.9	1.9	2.38
Oxides and hydroxides			
Brucite	−834.9	−925.9	63.1
Corundum	−1582.2	−1675.7	50.8
Diaspore	−920.8	−999.4	35.3
Hematite	−743.7	−825.6	87.4
Ilmenite	−1155.3	−1231.9	108.6
Magnetite	−1014.2	−1117.4	146.1
Periclase	−569.2	−601.5	27.0
Perovskite	−1575.3	−1660.6	93.6
Rutile	−889.5	−944.8	50.3
Spinel	−2176.5	−2300.3	84.5
Silica			
Coesite	−852.6	−907.6	39.4
Cristobalite, alpha	−853.9	−907.8	43.4
Cristobalite, beta	−853.3	−906.4	46.0
Quartz, alpha	−856.3	−910.7	41.5
Quartz, beta	−855.0	−908.6	44.2
Stishovite	−802.8	−861.3	27.8
Tridymite, low	−854.0	−907.8	43.8
Tridymite, high	−853.8	−907.1	45.5
Carbonates			
Aragonite	−1127.8	−1207.4	88.0
Calcite	−1128.3	−1206.8	91.7
Dolomite	−2162.4	−2325.3	154.9
Magnesite	−1029.9	−1113.6	65.2
Siderite	−666.7	−737.0	105.0

Table 8.8 (*cont.*)

Mineral	Free energy of formation $\Delta_f G$ (kJ mol^{-1})	Enthalpy of formation $\Delta_f H$ (kJ mol^{-1})	Entropy S (J mol^{-1} K^{-1})
Silicates			
Albite	−3703.3	−3291.6	224.4
Almandine	−4941.7	−5265.5	339.9
Andalusite	−2441.8	−2590.0	91.4
Anorthite	−4003.2	−4228.7	200.2
Cordierite	−8651.5	−9158.7	418.0
Diopside	−3026.2	−3200.6	142.5
Enstatite	−1458.6	−1545.9	66.3
Fayalite	−1379.4	−1479.4	148.3
Ferrosilite	−1117.5	−1194.4	95.9
Forsterite	−2055.0	−2174.4	94.0
Grossular	−6271.0	−6632.9	255.2
Jadeite	−2846.5	−3025.1	133.6
Kyanite	−2443.4	−2594.2	82.4
Sanidine	−3738.8	−3959.7	229.2
Microcline	−3745.4	−3970.8	214.2
Pyrope	−5936.0	−6286.6	266.4
Sillimanite	−2439.3	−2586.1	95.9
Titanite	−2455.1	−2596.7	129.3
Wollastonite	−1546.1	−1631.5	81.8
Pseudowollastonite	−1543.1	−1627.4	85.3
Zircon	−1918.9	−2033.4	84.0
Zoesite	−6494.2	−6889.5	297.6

Notes: The data for the Gibb's free energy and for the enthalpy are for formation of the mineral from the elements, at 1 bar and 298 K.

Sources: Berman, R. G. (1988). *Journal of Petrology*, **29**, 445–522. Robie, R. A. *et al.* (1978). Thermodynamic properties of minerals and related substances at 298.15 K and 1 Bar (10^5 Pascals) pressure and at higher temperatures. *Geological Survey Bulletin*, vol. 1452, Washington: DC, United States Government Printing Office.

Table 8.9 Radioactive isotopes: properties and applications

Z	Isotope	Abundance (%)	Decay constant[a] λ,a^{-1}	Half life[a] $t_{1/2}$, a	Main decay mode(s)	Products[a]	Applications
19	^{40}K	0.0117	4.962×10^{-10} 0.581×10^{-10} $\Sigma: 5.543 \times 10^{-10}$	1.397×10^9 11.93×10^9 $\Sigma: 1.25 \times 10^9$	$\beta-$ EC, $\beta+$	^{40}Ca (89%) ^{40}Ar (11%)	Petrogenetic tracer Geochronology, especially of K-bearing minerals and of high-grade metamorphics, cooling histories (^{40}Ar–^{39}Ar)
37	^{87}Rb	27.84	1.42×10^{-11}	48.8×10^9	$\beta-$, anti-neutrino	^{87}Sr	Geochronology, sea-water evolution and sediment correlation
57	^{138}La	0.09	$\sim2.3 \times 10^{-12}$	$\sim2.97 \times 10^{11}$	$\beta-$	^{138}Ce (34%)	Geochronology of ancient rocks; magma sources; ocean mixing
			4.44×10^{-12} $\Sigma: 6.75 \times 10^{-12}$	1.57×10^{11} $\Sigma: 1.03 \times 10^{11}$	EC	^{138}Ba (66%)	Geochronology of ancient rocks with lanthanum-rich minerals
62	^{147}Sm	15.0	6.54×10^{-12}	106×10^9	α	^{143}Nd	Geochronology, especially of Precambrian rocks and high-grade metamorphics; sediment provenance, crustal and mantle evolution, stony meteorite and lunar rock studies.

8

71	^{176}Lu	2.59	1.94×10^{-11}	35.7×10^9	$\beta-$; EC	^{176}**Hf** $(\sim 97\%)$; ^{176}Yb $(\sim 3\%)$	Geochronology, mantle evolution, crustal-growth modelling
75	^{187}Re	62.6	1.666×10^{-11}	41.6×10^9	$\beta-$	^{187}Os	Geochronology, including of iron meteorites; mantle and lithosphere evolution; petrogenesis; sea-water studies
78	^{190}Pt	0.013	1.477×10^{-12}	469×10^9	α	^{186}Os	Petrogenesis of Pt-bearing ores
90	^{232}Th	100	$0.494\,75 \times 10^{-10}$	14.01×10^9	α, γ	^{208}Pb, chain[b]	Geochronology, crustal evolution, meteorites
92	^{235}U	0.720	9.8485×10^{-10}	0.704×10^9	α, γ, F	^{207}Pb, chain[b]	Geochronology, crustal evolution, meteorites
92	^{238}U	99.2745	1.55125×10^{-10}	4.47×10^9	α, γ, F	^{206}Pb, chain[b]	Geochronology, crustal evolution, meteorites

Notes: a Decay constants and half lives are given for each of the decay modes where the radioisotope decays into more than one daughter product. The sum (Σ) for the overall decay is also given. The decay product that is used more widely in geochemistry is shown in bold. b The decay chain for each of these isotopes is given in Table 8.10, together with other data. EC = electron capture, F = spontaneous fission.

Sources: Dickin, A. P. (2005). *Radiogenic Isotope Geology*, 2nd edn. Cambridge: Cambridge University Press. Faure, G. and Mensing, T. M. (2005). *Isotope Principles and Applications*, 3rd edn. New Jersey: John Wiley. Parrington, J. R. et al. (1996). *Chart of the Nuclides*, 15th edn. General Electric Co. and KAPL Inc.

8

Table 8.10 Uranium and thorium decay series nuclides

	^{238}U			^{235}U			^{232}Th	
Nuclide	Decay mode	Half life	Nuclide	Decay mode	Half life	Nuclide	Decay mode	Half life
^{238}U	α	$4.4683 \pm 0.0048 \times 10^9$ a	^{235}U	α	$0.70381 \pm 0.00096 \times 10^9$ a	^{232}Th	α	$14.0 \pm 0.1 \times 10^9$ a
^{234}Th	β	24.1 days	^{231}Th	β	1.063 days	^{228}Ra	β	5.75 ± 0.03 a
^{234}Pa	β	6.69 hours	^{231}Pa	α	$32{,}760 \pm 220$ a	^{228}Ac	β	6.15 hours
^{234}U	α	$245\,250 \pm 490$ a	^{227}Ac	α, β	21.77 ± 0.02 a	^{228}Th	α	1.912 ± 0.002 a
^{230}Th	α	$75\,690 \pm 230$ a	^{227}Th	α	18.72 days	^{224}Ra	α	3.66 days
^{226}Ra	α	1599 ± 4 a	^{223}Fr	α, β	22 min	^{220}Rn	α	55.6 sec
^{222}Rn	α	3.823 ± 0.004 days	^{223}Ra	α	11.435 days	^{216}Po	α	0.145 sec
^{218}Po	α, β	3.04 min	^{219}At	α, β	50 sec	^{212}Pb	β	10.64 sec
^{218}At	α, β	1.6 sec	^{219}Rn	α	3.96 sec	^{212}Bi	α, β	1.009 hours
^{218}Rn	α	35 msec	^{215}Bi	β	7.7 min	^{212}Po	β	0.298 μsec
^{214}Pb	β	26.9 min	^{215}Po	α, β	1.78 msec	^{208}Tl	β	3.053 min
^{214}Bi	α, β	19.7 min	^{215}At	α	0.1 msec	^{208}Pb		Stable
^{214}Po	α	0.1637 msec	^{211}Pb	β	36.1 min			

8

^{210}Tl	β	1.3 min		^{211}Bi	α, β	2.14 min
^{210}Pb	α, β	22.6 ± 0.1 a		^{211}Po	α	0.516 sec
^{210}Bi	α, β	5.01 days		^{207}Tl	β	4.77 min
^{210}Po	α	138.4 ± 0.1 days		^{207}Pb		Stable
^{206}Hg	β	8.2 min				
^{206}Tl	β	4.2 min				
^{206}Pb		Stable				

Source: Based on a compilation in: Bourdon, B. et al. (2003). Introduction to uranium series geochemistry. In *Uranium Series Geochemistry*, ed. B. Bourdon et al., *Reviews in Mineralogy and Geochemistry*, vol. 52. Washington, DC: The Mineralogical Society of America. This gives references to the various sources of data.

8

Table 8.11 Chondrite Uniform Reservoir (CHUR) isotope ratio values

Ratio	Value
^{142}Nd/^{144}Nd	1.141 827
^{143}Nd/^{144}Nd	0.512 638
^{145}Nd/^{144}Nd	0.348417
^{146}Nd/^{144}Nd	0.721 9
^{148}Nd/^{144}Nd	0.241 578
^{150}Nd/^{144}Nd	0.236 418
^{147}Sm/^{144}Nd	0.196 6
^{176}Hf/^{177}Hf	0.282 86
^{176}Lu/^{177}Hf	0.033 4
^{187}Os/^{188}Os	0.128 63

Notes: The Chondrite Uniform Reservoir (CHUR) is the reference material based on chondrite meteorites against which the isotopic composition and evolution of other systems (e.g. the Earth's mantle) can be compared. All values above are those at the present day. They are the recommended values given by the sources below. The Nd isotope ratios are all normalised to ^{146}Nd/^{144}Nd = 0.7219.

Sources: Dickin, A. P. (2005). *Radiogenic Isotope Geology*, 2nd edn. Cambridge: Cambridge University Press. Faure, G. and Mensing, T. M. (2005). *Isotope Principles and Applications*, 3rd edn. New Jersey: John Wiley.

8

Table 8.12 Cosmogenic nuclides – production, properties and applications[a]

Z	Isotope	Production mode(s)	Main decay mode	Half life	Decay constant, a^{-1}	Application
1	^3H	$^{14}N(n,^3H)^{12}C$	β^- to ^3He	12.3 a	5.6×10^{-2}	Hydrology, groundwaters, glaciology, air–sea exchange
2	^3He	$^6Li(n,\alpha)^3H$, spallation	Stable	Stable	Stable	Groundwater studies, water tracer, rock exposure ages (<3000 years)
4	^7Be	Spallation	EC	53.28 days	4.747	Meteorology, air circulation, ocean water mixing (vertical)
4	^{10}Be	Spallation	β^-	1.52×10^6 a	4.56×10^{-7}	Dating ocean sediments, including Fe–Mn deposits (up to 10Ma), sedimentation and erosion rates, glaciation studies, igneous petrogenetic tracer, rock and meteorite exposure ages
6	^{14}C	$^{14}N(n,p)$	β^-	5730 a	1.209×10^{-4}	Archaeological dating (up to ~50 ka), carbonate deposit dating; ocean water tracer, palaeoclimates.
10	^{21}Ne	Spallation	Stable	Stable	Stable	Rock and meteorite exposure ages (long)
11	^{22}Na	Spallation	β^-, (EC)	2.604 a	2.66×10^{-1}	Meteorology, cloud physics, air circulation
13	^{26}Al	Spallation	β^-, (EC)	7.1×10^5 a	9.8×10^{-7}	Quartz-rich rock exposure ages (up to ~2.5 Ma), erosion rates
14	^{32}Si	Spallation	β^-	150 a	4.6×10^{-3}	Hydrology, groundwaters, glaciology, sediment chronology, biogeochemical cycles
15	^{32}P	Spallation	β^-	14.28 days	17.71	Meteorology, cloud physics, air circulation, biogeochemical cycles

(cont.)

8

Table 8.12 (*cont.*)

Z	Isotope	Production mode(s)	Main decay mode	Half life	Decay constant, a^{-1}	Application
	^{33}P	Spallation	β^-	25.3 days	9.998	Meteorology, cloud physics, air circulation, biogeochemical cycles
16	^{35}S	Spallation	β^-	87.2 days	2.901	Meteorology, cloud physics, air circulation
17	^{36}Cl	^{35}Cl(n,γ) spallation	β^-, (EC)	3.01×10^5 a	2.30×10^{-6}	Hydrological tracer; dating groundwaters; exposure ages and erosion rates including glacial erosion rates, meteorite terrestrial ages
18	^{39}Ar	Spallation, muon interaction	β^-	269 a	2.57×10^{-3}	Hydrology, groundwaters, glaciology, sediment chronology, vertical mixing in oceans
25	^{53}Mn	Spallation	EC	3.7×10^6 a	1.87×10^{-7}	Dating terrestrial and extra-terrestrial materials
36	^{81}Kr	Spallation	EC	2.1×10^5 a	3.30×10^{-6}	Sediment chronology, air–sea exchange, geochemical cycles, groundwater ages
53	^{129}I	Spallation	β^-	1.57×10^7 a	4.41×10^{-8}	Groundwater tracer
54	^{126}Xe	Spallation	Stable	Stable	Stable	Meteorite exposure ages

Notes: *a* Many nuclides are produced by cosmic ray interactions. The ones selected here are those more commonly used in the earth sciences. Cosmogenic nuclides are produced (in the atmosphere or surface materials) by three main routes: spallation of nuclides by high-energy neutrons, nuclear reaction by thermal neutron capture and muon induced nuclear disintegration.

Sources: Dickin, A. P. (2005). *Radiogenic Isotope Geology*, 2nd edn. Cambridge: Cambridge University Press. Faure, G. and Mensing, T. M. (2005). *Isotope Principles and Applications*, 3rd edn. New Jersey: John Wiley. Gosse, J. C. and Phillips, F. M. (2001). *Quaternary Science Reviews*, **20**, 1475–1560. Lal, D. (1988). *Annual Review of Earth and Planetary Science*, **16**, 355–388. Lal, D. (2003). Cosmogenic radionuclides. In *Encyclopedia of Atmospheric Sciences*, ed. J. R. Holton *et al.* Amsterdam: Academic Press.

Table 8.13 Extinct radionuclides – properties and applications

Z	Nuclide	Decay constant λ, a^{-1}	Half life $t_{1/2}$, a	Main decay mode(s)	Products	Applications
13	^{26}Al	9.76×10^{-7}	7.1×10^{5}	β^+, EC	^{26}Mg	Nucleosynthetic events, planetary differentiation
20	^{41}Ca	6.73×10^{-6}	1.03×10^{5}	EC	^{41}K	Nucleosynthetic events
25	^{53}Mn	1.9×10^{-7}	3.7×10^{6}	EC	^{53}Cr	Early Solar System development
26	^{60}Fe	4.6×10^{-7}	1.5×10^{6}	β	^{60}Ni	Early Solar System development, core formation in asteroids etc.
41	^{92}Nb	2×10^{-8}	3.5×10^{7}	EC	^{92}Zr	Nucleosynthesis events
46	^{107}Pd	1.1×10^{-7}	6.5×10^{6}	β	^{107}Ag	Early Solar System. Time gap between nucleosynthesis and Solar System formation
53	^{129}I	4.4×10^{-8}	1.57×10^{7}	β	^{129}Xe	Chondrite meteorite ages
62	^{146}Sm	6.7×10^{-9}	1.03×10^{8}	α	^{142}Nd	Early Earth differentiation, stages of nucleosynthesis
72	^{182}Hf	7.7×10^{-8}	9×10^{6}	β	^{182}W	Core formation in asteroids and planets
94	^{244}Pu	8.7×10^{-9}	8×10^{7}	EC, fission	Xe isotopes + others	Solar System origin
96	^{247}Cm	4.4×10^{-8}	1.56×10^{7}	α, β	^{235}U	Late stages of solar nebula

Notes: EC = electron capture

Sources: Dickin, A. P. (2005). *Radiogenic Isotope Geology*, 2nd edn. Cambridge: Cambridge University Press. Faure, G. and Mensing, T. M. (2005). *Isotope Principles and Applications*, 3rd edn. New Jersey: John Wiley. Parrington, J. R. et al. (1996). *Chart of the Nuclides*, 15th edn. General Electric Co. and KAPL Inc.

8

Table 8.14 Abundance and mass of naturally occurring nuclides

Z	Isotope	Abundance %	Mass[a]	Z	Isotope	Abundance %	Mass[a]
1	^{1}H	99.985	1.007 825		^{34}S	4.21	33.967 867
	^{2}H	0.015	2.0140		^{36}S	0.02	35.967 081
2	^{3}He	$\sim1.4 \times 10^{-4}$	3.016 03	17	^{35}Cl	75.77	34.968 853
	^{4}He	~100	4.002 60		^{37}Cl	24.23	36.965 903
3	^{6}Li	7.5	6.015 123	18	^{36}Ar	0.34	35.967 545
	^{7}Li	92.5	7.016 004		^{38}Ar	0.06	37.962 732
4	^{9}Be	100	9.012 182		^{40}Ar	99.60	39.962 383
5	^{10}B	19.9	10.012 937	19	^{39}K	93.2581	38.963 707
	^{11}B	80.1	11.009 305		^{40}K	0.0117	39.963 999
6	^{12}C	98.90	12.000 000		^{41}K	6.7302	40.961 826
	^{13}C	1.10	13.003 355	20	^{40}Ca	96.941	39.962 591
7	^{14}N	99.63	14.003 074		^{42}Ca	0.647	41.958 618
	^{15}N	0.37	15.000 109		^{43}Ca	0.135	42.958 767
8	^{16}O	99.76	15.994 915		^{44}Ca	2.086	43.955 482
	^{17}O	0.04	16.999 132		^{46}Ca	0.004	45.953 693
	^{18}O	0.20	17.999 161		^{48}Ca	0.187	47.952 534
9	^{19}F	100	18.998 403	21	^{45}Sc	100	44.955 912
10	^{20}Ne	90.50	19.992 440	22	^{46}Ti	8.25	45.952 632
	^{21}Ne	0.27	20.993 847		^{47}Ti	7.44	46.951 763
	^{22}Ne	9.23	21.991 385		^{48}Ti	73.72	47.947 946
11	^{23}Na	100	22.989 769		^{49}Ti	5.41	48.947 870
12	^{24}Mg	78.99	23.985 042		^{50}Ti	5.18	49.944 791
	^{25}Mg	10.00	24.985 837	23	^{50}V	0.250	49.947 159
	^{26}Mg	11.01	25.982 593		^{51}V	99.750	50.943 960
13	^{27}Al	100	26.981 539	24	^{50}Cr	4.35	49.946 044
14	^{28}Si	92.23	27.976 927		^{52}Cr	83.79	51.940 508
	^{29}Si	4.67	28.976 495		^{53}Cr	9.50	52.940 649
	^{30}Si	3.10	29.973 770		^{54}Cr	2.36	53.938 880
15	^{31}P	100	30.973 762	25	^{55}Mn	100	54.938 045
16	^{32}S	95.02	31.972 071	26	^{54}Fe	5.85	53.939 611
	^{33}S	0.75	32.971 459		^{56}Fe	91.75	55.934 938

8

Table 8.14 (*cont.*)

Z	Isotope	Abundance %	Mass[a]	Z	Isotope	Abundance %	Mass[a]
	^{57}Fe	2.12	56.935 394	36	^{78}Kr	0.35	77.920 365
	^{58}Fe	0.28	57.933 276		^{80}Kr	2.26	79.916 379
27	^{59}Co	100	58.933 195		^{82}Kr	11.52	81.913 484
28	^{58}Ni	68.08	57.935 343		^{83}Kr	11.47	82.914 136
	^{60}Ni	26.22	59.930 786		^{84}Kr	57.0	83.911 507
	^{61}Ni	1.14	60.931 056		^{86}Kr	17.4	85.911 611
	^{62}Ni	3.63	61.928 345	37	^{85}Rb	72.16	84.911 790
	^{64}Ni	0.93	63.927 966		^{87}Rb	27.84	86.909 181
29	^{63}Cu	69.17	62.939 598	38	^{84}Sr	0.56	83.913 425
	^{65}Cu	30.83	64.927 790		^{86}Sr	9.86	85.909 260
30	^{64}Zn	48.6	63.929 142		^{87}Sr	7.00	86.908 877
	^{66}Zn	27.9	65.926 033		^{88}Sr	82.58	87.905 612
	^{67}Zn	4.1	66.927 127	39	^{89}Y	100	88.905 848
	^{68}Zn	18.8	67.924 844	40	^{90}Zr	51.45	89.904 704
	^{70}Zn	0.6	69.925 319		^{91}Zr	11.22	90.905 646
31	^{69}Ga	60.11	68.925 574		^{92}Zr	17.15	91.905 041
	^{71}Ga	39.89	70.924 701		^{94}Zr	17.38	93.906 315
32	^{70}Ge	21.23	69.924 247		^{96}Zr	2.80	95.908 273
	^{72}Ge	27.66	71.922 076	41	^{93}Nb	100	92.906 378
	^{73}Ge	7.73	72.923 459	42	^{92}Mo	14.84	91.906 811
	^{74}Ge	35.94	73.921 178		^{94}Mo	9.25	93.905 088
	^{76}Ge	7.44	75.921 403		^{95}Mo	15.92	94.905 842
33	^{75}As	100	74.921 596		^{96}Mo	16.68	95.904 680
34	^{74}Se	0.89	73.922 476		^{97}Mo	9.55	96.906 022
	^{76}Se	9.36	75.919 214		^{98}Mo	24.13	97.905 408
	^{77}Se	7.63	76.919 914		^{100}Mo	9.63	99.907 477
	^{78}Se	23.78	77.917 309	44	^{96}Ru	5.52	95.907 599
	^{80}Se	49.61	79.916 521		^{98}Ru	1.88	97.905 287
	^{82}Se	8.73	81.916 699		^{99}Ru	12.7	98.905 939
35	^{79}Br	50.69	78.918 337		^{100}Ru	12.6	99.904 219
	^{81}Br	49.31	80.916 291		^{101}Ru	17.0	100.905 582

(*cont.*)

Table 8.14 (*cont.*)

Z	Isotope	Abundance %	Mass[a]	Z	Isotope	Abundance %	Mass[a]
	^{102}Ru	31.6	101.904 349	51	^{121}Sb	57.3	120.903 816
	^{104}Ru	18.7	103.905 433		^{123}Sb	42.7	122.904 214
45	^{103}Rh	100	102.905 504	52	^{120}Te	0.096	119.904 020
46	^{102}Pd	1.02	101.905 609		^{122}Te	2.600	121.903 044
	^{104}Pd	11.14	103.904 036		^{123}Te	0.908	122.904 270
	^{105}Pd	22.33	104.905 085		^{124}Te	4.816	123.902 818
	^{106}Pd	27.33	105.903 486		^{125}Te	7.14	124.904 431
	^{108}Pd	26.46	107.903 892		^{126}Te	18.95	125.903 312
	^{110}Pd	11.72	109.905 153		^{128}Te	31.69	127.904 463
47	^{107}Ag	51.839	106.905 097		^{130}Te	33.8	129.906 224
	^{107}Ag	48.161	108.904 752	53	^{127}I	100	126.904 473
48	^{106}Cd	1.25	107.906 460	54	^{124}Xe	0.10	123.905 893
	^{108}Cd	0.89	107.904 184		^{126}Xe	0.09	125.904 274
	^{110}Cd	12.49	109.903 002		^{128}Xe	1.92	127.903 531
	^{111}Cd	12.80	110.904 178		^{129}Xe	26.4	128.904 779
	^{112}Cd	24.13	111.902 758		^{130}Xe	4.1	129.903 508
	^{113}Cd	12.22	112.904 401		^{131}Xe	21.2	130.905 082
	^{114}Cd	28.73	113.903 359		^{132}Xe	26.9	131.904 154
	^{116}Cd	7.49	115.904 756		^{134}Xe	10.4	133.905 395
49	^{113}In	4.29	112.904 058		^{136}Xe	8.9	135.907 219
	^{115}In	95.71	114.903 878	55	^{133}Cs	100	132.905 452
50	^{112}Sn	0.97	111.904 818	56	^{130}Ba	0.106	129.906 321
	^{114}Sn	0.65	113.902 779		^{132}Ba	0.101	131.905 061
	^{115}Sn	0.34	114.903 342		^{134}Ba	2.42	133.904 508
	^{116}Sn	14.54	115.901 741		^{135}Ba	6.593	134.905 689
	^{117}Sn	7.68	116.902 952		^{136}Ba	7.85	135.904 576
	^{118}Sn	24.22	117.901 603		^{137}Ba	11.23	136.905 827
	^{119}Sn	8.59	118.903 308		^{138}Ba	71.70	137.905 247
	^{120}Sn	32.59	119.902 195	57	^{138}La	0.090	137.907 112
	^{122}Sn	4.63	121.903 439		^{139}La	99.910	138.906 353
	^{124}Sn	5.79	123.905 274	58	^{136}Ce	0.19	135.907 172

Table 8.14 (*cont.*)

Z	Isotope	Abundance %	Mass[a]	Z	Isotope	Abundance %	Mass[a]
	^{138}Ce	0.25	137.905 991		^{161}Dy	18.9	160.926 933
	^{140}Ce	88.48	139.905 439		^{162}Dy	25.5	161.926 798
	^{142}Ce	11.08	141.909 244		^{163}Dy	24.9	162.928 731
59	^{141}Pr	100	140.907 653		^{164}Dy	28.2	163.929 175
60	^{142}Nd	27.13	141.907 723	67	^{165}Ho	100	164.930 322
	^{143}Nd	12.18	142.909 814	68	^{162}Er	0.14	161.928 778
	^{144}Nd	23.80	143.910 087		^{164}Er	1.61	163.929 200
	^{145}Nd	8.30	144.912 574		^{166}Er	33.6	165.930 293
	^{146}Nd	17.19	145.913 117		^{167}Er	22.95	166.932 048
	^{148}Nd	5.76	147.916 893		^{168}Er	26.8	167.932 370
	^{150}Nd	5.64	149.920 891		^{170}Er	14.9	169.935 464
62	^{144}Sm	3.1	143.911 999	69	^{169}Tm	100	168.934 213
	^{147}Sm	15.0	146.914 898	70	^{168}Yb	0.13	167.933 897
	^{148}Sm	11.3	147.914 823		^{170}Yb	3.05	169.934 762
	^{149}Sm	13.8	148.917 185		^{171}Yb	14.3	170.936 326
	^{150}Sm	7.4	149.917 276		^{172}Yb	21.9	171.936 382
	^{152}Sm	26.7	151.919 732		^{173}Yb	16.12	172.938 211
	^{154}Sm	22.7	153.922 209		^{174}Yb	31.8	173.938 862
63	^{151}Eu	47.8	150.919 850		^{176}Yb	12.7	175.942 572
	^{153}Eu	52.2	152.921 230	71	^{175}Lu	97.41	174.940 772
64	^{152}Gd	0.20	151.919 791		^{176}Lu	2.59	175.942 686
	^{154}Gd	2.18	153.920 866	72	^{174}Hf	0.162	173.940 046
	^{155}Gd	14.80	154.922 622		^{176}Hf	5.206	175.941 409
	^{156}Gd	20.47	155.922 123		^{177}Hf	18.606	176.943 221
	^{157}Gd	15.65	156.923 960		^{178}Hf	27.297	177.943 699
	^{158}Gd	24.84	157.924 104		^{179}Hf	13.629	178.945 816
	^{160}Gd	21.86	159.927 054		^{180}Hf	35.100	179.946 550
65	^{159}Tb	100	158.925 347	73	^{180}Ta	0.012	179.947 465
66	^{156}Dy	0.06	155.924 283		^{181}Ta	99.988	180.947 996
	^{158}Dy	0.10	157.924 409	74	^{180}W	0.120	179.946 704
	^{160}Dy	2.34	159.925 198		^{182}W	26.498	181.948 204

(*cont.*)

Table 8.14 (*cont.*)

Z	Isotope	Abundance %	Mass[a]	Z	Isotope	Abundance %	Mass[a]
	^{183}W	14.314	182.950 223	79	^{190}Au	100	196.966 569
	^{184}W	30.642	183.950 931	80	^{196}Hg	0.15	195.965 833
	^{186}W	28.426	185.954 364		^{198}Hg	9.97	197.966 769
75	^{185}Re	37.40	184.952 950		^{199}Hg	16.87	198.968 280
	^{187}Re	62.60	186.955 753		^{200}Hg	23.10	199.968 326
76	^{184}Os	0.020	183.952 489		^{201}Hg	13.18	200.970 302
	^{186}Os	1.58	185.953 838		^{202}Hg	29.86	201.970 643
	^{187}Os	1.6	186.955 751		^{204}Hg	6.87	203.973 494
	^{188}Os	13.3	187.955 838	81	^{203}Tl	29.52	202.972 344
	^{189}Os	16.1	188.958 148		^{205}Tl	70.48	204.974 428
	^{190}Os	26.4	189.958 447	82	^{204}Pb	1.4	203.973 044
	^{192}Os	41.0	191.961 481		^{206}Pb	24.1	205.974 465
77	^{191}Ir	37.3	190.960 594		^{207}Pb	22.1	206.975 897
	^{193}Ir	62.7	192.962 926		^{208}Pb	52.4	207.976 652
78	^{190}Pt	0.01	189.959 932	83	^{209}Bi	100	208.980 399
	^{192}Pt	0.79	191.961 038	90	^{232}Th	100	232.038 055
	^{194}Pt	32.9	193.962 680	92	^{234}U	0.0055	234.040 952
	^{195}Pt	33.8	194.964 791		^{235}U	0.720	235.043 930
	^{196}Pt	25.3	195.964 952		^{238}U	99.2745	238.050 788
	^{198}Pt	7.2	197.967 893				

Notes: *a* Atomic mass units based on the ^{12}C standard. Radioactive nuclides are italicised. Data for other radionuclides can be found in Tables 8.9 and 8.10, for cosmogenic nuclides in Table 8.12 and for inert gases in Table 5.3.

Sources: Masses: Lide, D. R., ed. (2008). *Handbook of Chemistry and Physics, 89th. edn, 2008–2009.* Boca Raton, IL: CRC Press. Abundances: Parrington, J. R. *et al.* (1996). *Chart of the Nuclides.* General Electric Co. and KAPL Inc. See also Table 5.3 for sources and more details on inert gases.

Table 8.15 Stable isotope notation

δ *notation*
$$\delta_x\text{‰} = \left(\frac{R_x - R_{std}}{R_{std}}\right) 10^3$$

where R_x is the isotopic ratio (e.g. $^{18}O/^{16}O$) of the sample x, and R_{std} that of the reference standard (e.g. SMOW). The δ value is in per mil (‰).

ε *notation*
$$\varepsilon^{mm}R = \left(\frac{R_x - R_{std}}{R_{std}}\right) 10^4$$

where ^{mm}R (e.g. ^{18}O) is an identifier for the isotope ratio being measured (e.g. $^{18}O/^{16}O$); R_x etc. as above. Note that the ε value is given in parts per 10 000.

Fractionation factor (α)
$$\alpha_{A-B} = \frac{R_A}{R_B} = \frac{1000 + \delta_A}{1000 + \delta_B}$$

where R_A and R_B are coexisting materials (e.g. minerals).

δ *relationship*
$$\Delta_{A-B} = \delta_A - \delta_B \approx 10^3 \ln \alpha_{A-B}$$
$$\text{for values of } \delta \leq 10.$$

Note: This is the standard notation used in stable isotope studies and as cited by texts and papers on the subject. It is a convenient way of presenting the small variations observed in isotopic ratios. The overall abundance ratio values for commonly studied stable isotopes are given in Table 8.16 and ratios in reference materials in Table 8.18.

8

Table 8.16 Stable isotope ratio values and example applications

Z	Isotope ratio	Abundance ratio value	Applications
1	$^2H/^1H$	0.000 15	Water–rock interactions. Magma volatiles and degassing. Biochemical processes. See also O with H.
3	$^7Li/^6Li$	12.333 33	Studies of chemical weathering and mineral precipitation. Tracer of crustal material in mantle.
5	$^{11}B/^{10}B$	4.025 126	Past seawater pH. Meteorite studies.
6	$^{13}C/^{12}C$	0.011 122 3	Numerous, including: Composition of early atmosphere, early life, meteorite studies, mantle heterogeneity, origins of diamond.
7	$^{15}N/^{14}N$	0.003 713 7	Ocean nitrate utilisation. Mixing of waters, especially fresh with seawater.
8	$^{18}O/^{16}O$	0.000 200 48	Numerous, including palaeothermometry (especially of carbonate-bearing systems), igneous and metamorphic rock geothermometer, meteorite and extraterrestrial rock studies, water–rock interaction, sediment and marine mineral origins, water provenance, brine evolution.
8 with 1	$^{18}O/^{16}O$ with $^2H/^1H$		Numerous, including past climatic conditions, water–rock interactions (including hydrothermal systems), water mixing, origin of brines.
12	$^{25}Mg/^{24}Mg$ or $^{26}Mg/^{24}Mg$	0.126 8 or 0.139 6	Meteorite studies.

Table 8.16 (*cont.*)

Z	Isotope ratio	Abundance ratio value	Applications
14	$^{30}Si/^{28}Si$	0.033 611 6	Weathering intensity. Ocean Si utilisation. Secondary mineral formation. Composition of the Earth's core.
16	$^{34}S/^{32}S$	0.044 306 4	Role of sulphate-reducing bacteria in formation of sulphate deposits. Temperatures and conditions of formation of some sulphide ores. Sulphur cycle.
17	$^{37}Cl/^{35}Cl$	0.319 783 5	Aerosol studies, especially involving oceans and atmosphere. Origin of salts in waters.
20	$^{44}Ca/^{42}Ca$ or $^{44}Ca/^{40}Ca$	3.224 or 0.002 152	Ocean Ca cycle, palaeothermometer, tracer of organic processes.
26	$^{56}Fe/^{54}Fe$	15.684	Biological activity. Terrestrial redox processes.
42	$^{97}Mo/^{95}Mo$	0.599 874 3	Iron–manganese ocean deposits. Past-oceanic redox conditions.

Sources: Based on: Faure, G. and Mensing, T. M. (2005). *Isotope Principles and Applications*, 3rd edn. New Jersey: John Wiley. Hoefs, J. (1997). *Stable Isotope Geochemistry*. Berlin: Springer-Verlag. Johnson, C. M. *et al.* (2004). Geochemistry of non-traditional stable isotopes. In *Reviews in Mineralogy and Geochemistry*, **55**. Washington, DC: Mineralogical Society of America, Geochemical Society. Sharp, Z. (2007). *Principles of Stable Isotope Geochemistry*. New Jersey: Pearson, Prentice Hall.

8

Table 8.17 Stable isotope ranges for $\delta^{11}B$, $\delta^{13}C$, $\delta^{15}N$, $\delta^{18}O$ in natural materials

Material	Range	
	From	To
$\delta^{11}B$ *relative to standard SRM951*		
MORB	~+2	
Igneous rocks	−16	+8
Metamorphic rocks	−34	+23
Sediments	−17	+28
Evaporites	−32	+35
Marine carbonates	0	+30
Ocean water	~+40	
$\delta^{13}C$ *relative to standard PDB*		
Chondritic meteorites	−30	−12
MORB	−10	0
Metamorphic and igneous carbonate	−12	+25
Marine carbonates	−6	+4
Freshwater carbonates	−15	+5
Marine sediments (organic carbon)	−130	+7
Lake sediments	−35	−10
Coal	−30	−18
Oil	−36	−15
Diamond	−35	+5
Plants, land	−35	−10
Organisms, marine	−74	−2
Volcanic methane	−36	−20
Natural gas methane	−70	−20
Volcanic carbon dioxide	−36	+28
Atmospheric carbon dioxide	~−8	
$\delta^{15}N$ *relative to standard AIR*		
Meteorites	0	+50
MORB	−10	+5
Igneous rocks	−35	+30

Table 8.17 (*cont.*)

Material	Range	
	From	To
δ¹⁵N relative to standard AIR		
Metamorphic rocks	0	+28
Diamond	−38	+12
Ground waters	−1	+6
Ocean water	−4	+18
Volcanic gases/hot springs	−10	+15
Plants	−12	+15
Soils and organic material	−11	+16
δ¹⁸O relative to standard SMOW		
MORB	+5	+6
Basalt	+3	+11
Andesite	+5	+10
Granite	−5	+14
Metamorphic rocks	+2	+28
Sedimentary rocks	+2	+40
Marine limestone	+18	+34
Plants	0	+40
Atmospheric carbon dioxide	+40	+50

The table uses $\delta^{15}N$ relative to standard AIR and $\delta^{18}O$ relative to standard SMOW.

Notes: see Fig. 1.1 for data on oxygen isotopes in meteorites, Table 4.14 and Fig. 4.3 for oxygen and hydrogen isotopes in natural waters, and Figs. 7.3–7.5 for oxygen and carbon isotope variations with geological time.

Sources: Coplen, T. B. *et al.* (2002). Compilation of minimum and maximum isotope ratios of selected elements in naturally occurring terrestrial materials and reagents. *USGS Water-Resources Investigation Report,* **01-4222**. Reston, VA: US Department of the Interior, USGS. (This report also provides some data on other isotopes, see also http://pubs.water.usgs.gov/wri014222.) Dawson, T. E. and Siegwolf, R. T. W., eds. (2007). *Stable Isotopes as Indicators of Ecological Change.* Amsterdam: Elsevier-Academic Press. Hoefs, J. (1977). *Stable Isotope Geochemistry.* Berlin: Springer-Verlag. Sharp, Z. (2007). *Principles of Stable Isotope Geochemistry.* New Jersey: Pearson, Prentice Hall.

8

Table 8.18 Selected stable isotope reference standards

Standard name	Material	δD (SMOW)	$\delta^{13}C$ (PDB)	$\delta^{15}N$ (AIR)	$\delta^{18}O$ (PDB)	$\delta^{18}O$ (SMOW)
VSMOW	Vienna Standard Mean Ocean Water	0.00				0.00
SLAP	Standard Light Antarctic Precipitation (water)	−428				−55.50
GISP	Greenland Ice Sheet Precipitation (water)	−189.73				−24.78
NBS-14	N_2 gas			−1.18		
NBS-18	Calcite		−5.014		−23.2	7.2
NBS-19	Calcite		1.95		−2.20	28.64
NBS-28	Silica sand					9.57
NBS-30	Biotite	−65.7				5.12
IAEA-CO-1	Calcite		2.48		−2.44	28.39
IAEA-CO-8	Calcite		−5.75		−22.67	7.54
IAEA-CO-9	$BaCO_3$		−47.3		−15.6	15.16
IAEA-N-1	$(NH_4)_2SO_4$			0.43		
IAEA-N-2	$(NH_4)_2SO_4$			20.32		
IAEA-NO-3	KNO_3			4.69		25.6
NSVEC	N_2 gas			−2.77		
LSVEC	Li_2CO_3		−46.48		−26.7	3.63
USGS24	Graphite		−15.99			
USGS25	$(NH_4)_2SO_4$			−30.4		
USGS26	$(NH_4)_2SO_4$			53.7		
USGS32	KNO_3			180		25.7
USGS35	$NaNO_3$			2.7		57.5

Table 8.18 (*cont.*)

Standard name	Material	$\delta^{11}B$ (SRM951)	$\delta^{18}O$ (SMOW)	$\delta^{34}S$ (CDT)	$\delta^{37}Cl$ (SMOC)
IAEA-S-1	Ag_2S			−0.30	
IAEA-S-2	Ag_2S			22.67	
IAEA-S-3	Ag_2S			−32.3	
IAEA-SO-5	$BaSO_4$		12.0	0.49	
IAEA-SO-6	$BaSO_4$		−11.0	−34.05	
NBS-123	sphalerite			17.44	
NBS-127	$BaSO_4$		8.7	21.1	
SRM 975	NaCl				0.43
SRM 975a	NaCl				0.2
ISL 354	NaCl				0.05
IAEA-B-1	Seawater	38.6			
IAEA-B-2	Groundwater	13.8			
IAEA-B-3	Tourmaline	−8.7			

Note: This table gives data on several of the standards in use in isotope studies, and in relation to the initial principal reference standards: SMOW: Standard Mean Ocean Water, PDB: Pee Dee Belemnite, CDT: Canyon Diablo Troilite, AIR: Air, SMOC: Standard Mean Ocean Chloride. SRM951 is boric acid. The latest information on isotope standards can be found on the IAEA organisation and National Institute of Standards and Technology websites.

Sources: International Atomic Agency: http://www.iaea.org/programme/aqcs/pdf/catalogue.pdf
National Institute of Standards and Technology:
http://ts.nist.gov/measurementservices/referencematerials/index.cfm accessed August 2008.
Sharp, Z. (2007). *Principles of Stable Isotope Geochemistry.* New Jersey: Pearson, Prentice Hall.

∞

9 Crystallography and mineralogy

This chapter contains basic information on crystallography of relevance to earth scientists, and properties of common minerals. Further information on minerals can be found in several other chapters, especially Chapters 2, 3 and 10.

Table 9.1 The seven crystal systems

System	Minimum defining symmetry[*]	Cell axes	Axes angles
Triclinic	none	$a \neq b \neq c$	$\alpha \neq \beta \neq \gamma \neq 90°$
Monoclinic	1 diad axis	$a \neq b \neq c$	$\alpha = \gamma = 90° \neq \beta$
Orthorhombic	3 diad axes	$a \neq b \neq c$	$\alpha = \beta = \gamma = 90°$
Tetragonal	1 tetrad axis	$a = b \neq c$	$\alpha = \beta = \gamma = 90°$
Cubic	4 triad axes	$a = b = c$	$\alpha = \beta = \gamma = 90°$
Trigonal	1 triad axis	$a = b = c$	$\alpha = \beta = \gamma \neq 90°$
		$a = b \neq c$	$\alpha = \beta = 90° \gamma = 120°$
Hexagonal	1 hexad axis	$a = b \neq c$	$\alpha = \beta = 90° \gamma = 120°$

Note: [*] In terms of rotation axes of symmetry, which may be inversion axes.

Source: Hahn, Th., ed. (2002). *International Tables for Crystallography. Brief Teaching Edition of Volume A Space Group Symmetry*. Dordrecht: Kluwer Academic Publishers.

9

Table 9.2 The thirty-two crystal classes

System	Class Symbol	Symmetry					
		Axes[a]				Planes	Centre
		2	3	4	6		
Triclinic	1	−	−	−	−	−	−
	$\bar{1}$	−	−	−	−	−	*
Monoclinic	m	−	−	−	−	1	−
	2	1	−	−	−	−	−
	2/m	1	−	−	−	1	*
Orthorhombic	$mm2$	1	−	−	−	2	−
	222	3	−	−	−	−	−
	mmm	3	−	−	−	3	*
Trigonal	3	−	1	−	−	−	−
	$\bar{3}$	−	1i	−	−	−	−
	3m	−	1	−	−	3	−
	$\bar{3}m$	3	1i	−	−	3	−
	32	3	1	−	−	−	−
Tetragonal	4	−	−	1	−	−	−
	$\bar{4}$	−	−	1i	−	−	−
	4/m	−	−	1	−	1	*
	4mm	−	−	1	−	4	−
	$\bar{4}2m$	2	−	1i	−	2	−
	422	4	−	1	−	−	−
	4/mmm	4	−	1	−	5	*
Hexagonal	6	−	−	−	1	−	−
	$\bar{6}$	−	−	−	1i	−	−
	6/m	−	−	−	1	1	*
	6mm	−	−	−	1	6	−
	$\bar{6}2m$	−	−	−	1i	3	−
	622	6	−	−	1	−	−
	6/mmm	6	−	−	1	7	*

Table 9.2 (*cont.*)

System	Class Symbol	2	3	4	6	Planes	Centre
		Symmetry Axes[a]					
Cubic	23	3	4	−	−	−	−
	$m\bar{3}$	3	4	−	−	3	*
	$\bar{4}3m$	−	4	$3i$	−	6	−
	432	6	4	3	−	−	−
	$m\bar{3}m$	6	4	3	−	9	*

Note: a The axes are of rotational symmetry, the numbers referring to the degree of the axis: 2 = diad, 3 = triad etc. *i* indicates inversion axis or axes. Inversion axes can be expressed in other symmetry elements, e.g. an inversion hexad axis is equivalent to a triad axis normal to a plane of symmetry.

Source: As for Table 9.1.

Table 9.3 Radius ratio of ions and co-ordination

R_A/R_B^a	No. B ions around A (maximum)	Nature of co-ordination
~1.0	12	Cubic or hexagonal close-packing
1.00–0.73	8	Corners of cube
0.73–0.41	6	Corners of octahedron
	4	Corners of square
0.41–0.22	4	Corners of tetrahedron
0.22–0.15	3	Corners of triangle
<0.15	2	Linear

Note: a R_A/R_B is the ratio of the size of the A ionic radius over that of the B ion, where A is the smaller ion. The ratio of the radii of adjoining ions can be a useful guide in indicating the number and arrangement of nearest neighbours. It is only approximate since it is based on a 'hard sphere' model.

Table 9.4 Structural classification of silicates

Name	Synonyms	Arrangement of Si–O tetrahedra	Si:O atom ratio	Mineral group example	Mineral examples[a]
Island[b]	Nesosilicate	Independent	1:4	Olivine	Forsterite Titanite
	Sorosilicate	Two sharing one oxygen	2:7	Epidote	Clinozoisite Lawsonite
Ring	Cyclosilicate	Closed rings	1:3		Beryl Tourmaline
Chain	Inosilicate	Single chains	1:3	Pyroxene	Augite
		Double chains	4:11	Amphibole	Hornblende
Sheet	Phyllosilicate	Continuous sheets	2:5	Mica	Biotite Kaolinite
Framework	Tectosilicate	Continuous framework	1:2	Feldspar	Albite Quartz

Notes: a The mineral examples are not necessarily members of the group examples.
b The term 'orthosilicate' is sometimes used.

Source: This classification is used in many sources, e.g. Klein, C. (2002). *The Manual of Mineral Science*, 22nd edn. New York: Wiley.

Table 9.5 Chemical classification of non-silicate minerals

Class[a]	Common formulae[b]	Mineral examples Name	Mineral examples Formula
Native elements and alloys	X	Diamond	C
Sulphides inc. sulpharsenides and arsenides	XS, XAsS, XAs	Cinnabar Arsenopyrite Nickeline	HgS FeAsS NiAs
Sulphosalts[c]	Various	Energite Tetrahedrite	Cu_3AsS_4 $Cu_{12}Sb_4S_{13}$
Oxides and hydroxides	Various inc. XO, XY_2O_4, $X(OH)_2$	Rutile Brucite	TiO_2 $Mg(OH)_2$
Halides	XH, XH_2	Halite Fluorite	NaCl CaF_2
Carbonates	XCO_3, $XY(CO_3)_2$	Aragonite Malachite	$CaCO_3$ $Cu_2CO_3(OH)_2$
Nitrates	XNO_3	Nitratine	$NaNO_3$
Borates	$X_mB_4O_6(OH)_n$	Borax	$Na_2B_4O_5(OH)_4.8H_2O$
Niobates and tantalates	$XNbO_3$	Lueshite	$NaNbO_3$
Phosphates, arsenates and vanadates	Various	Monazite Olivenite Hendersonite	$(Ce,La,Nd,Th)PO_4$ $Cu_2AsO_4.OH$ $Ca_2V_9O_{24}.8H_2O$
Sulphates	Various	Celestine	$SrSO_4$
Chromates, tungstates and molybdates	$XCrO4$, $(X,Y)WO_4$, $XMoO_4$	Crocoite Scheelite Wulfenite	$PbCrO_4$ $CaWO_4$ $PbMoO_4$

Notes and Sources: a Some classes contain particular mineral groups, e.g. the carbonate division contains the aragonite, carbonate and dolomite groups. b X and Y stand for elements usually but not exclusively metals, H for a halide. Other letters have their usual chemical meaning. c Sulphosalts are structurally distinct from sulphides.

There is no generally accepted 'standard' classification of non-silicate minerals. The above chemical one is an adaptation of that initially given by Dana and Dana and is based on: Clark, A. M. (1993). *Hey's Mineral Index*, 3rd edn. London: Chapman and Hall. Dana, J. D. and Dana, E. S. (1944–1962). *The System of Mineralogy*, 7th edn. 3 volumes. Rewritten by Palache, C., Herman, H. and Frondel, C. New York: John Wiley. Klein, C. (2002). *The Manual of Mineral Sciences*, 22nd edn. New York: John Wiley. It does not cover all the rare minerals, such as carbides and oxysulphides, which are sometimes given their own class. Clark gives a full list of minerals under each of the chemical sections in his work.

Table 9.6 Common silicate and silica minerals

Mineral	Formula	CS^a	OS^b	D^c	H^d	Struc.^{e}	Group
Aegirine	$NaFeSi_2O_6$	M	–	3.5–3.6	6	Chain	Pyroxene
Aegirine-augite	$(Na,Ca)(Fe,Mg,Al)Si_2O_6$	M	–/+	3.4–3.6	6	Chain	Pyroxene
Aenigmatite	$NaFe_5TiSi_6O_{20}$	Tric	–	3.74–3.86	5.5–6	Chain	Pyroxene
Åkermanite	$Ca_2MgSi_2O_7$	Tet	+	2.94	5–6	Island	Melilite
Albite	$NaAlSi_3O_8$	Tric	+	2.62	6–6.5	Fmwk	Feldspar
Allanite	$(Ce,Ca,Y)_2(Al,Fe)_3(SiO_4)_3OH$	M	–/+	3.4–4.2	5–6.5	Island	Epidote
Almandine	$Fe_3^{2+}Al_2(SiO_4)_3$	C		4.318	7	Island	Garnet
Analcime	$NaAlSi_2O_6.H_2O$	C		2.24–2.29	5.5	Fmwk	
Andalusite	Al_2SiO_5	O	–	3.14–3.16	6.5–7.5	Island	
Andradite	$Ca_3Fe_2^{3+}(SiO_4)_3$	C		3.859	7	Island	Garnet
Anorthite	$CaAl_2Si_2O_8$	Tric	–	2.76	6–6.5	Fmwk	Feldspar
Anorthoclase	$(Na,K)AlSi_3O_8$	Tric	–	2.56–2.62	6	Fmwk	Feldspar
Anthophyllite	$(Mg,Fe)_7Si_8O_{22}(OH)_2$	O	–/+	2.9–3.5	5.5–6	Chain	Amphibole
Antigorite	$(Mg,Fe)_3Si_2O_5(OH)_4$	M	–	2.6	2.5–3.5	Sheet	Kyanite–serpentine
Arfvedsonite	$Na_3Fe_4^{2+}Fe^{3+}Si_8O_{22}(OH)_2$	M	–	3.30–3.50	5–6	Chain	Amphibole
Augite	$(Ca,Mg,Fe)_2Si_2O_6$	M	+	3.19–3.56	5.5–6	Chain	Pyroxene
Axinite	$(Ca,Mn,Fe)Al_2BSi_4O_{15}(OH)$	Tric	–	3.18–3.43	6.5–7	Ring	

6

Name	Formula	CS	OS	D	H	Struc.	Group
Beryl	$Be_3Al_2Si_6O_{18}$	H	—	2.66–2.92	7.5–8	Ring	
Biotite	$K(Mg,Fe^{2+})_3(Al,Fe^{3+})Si_3O_{10}(OH,F)_2$	M	—	2–3	2.5–3	Sheet	Mica
Bustamite	$(Mn,Ca)SiO_3$	Tric	—	3.32–3.43	5.5–6.5	Chain	
Cancrinite	$Na_6Ca_2Al_6Si_6O_{24}(CO_3)_2$	H	—	2.4–2.5	5–6	Fmwk	Cancrinite
Chabazite	$CaAl_2SiO_4 \cdot 6H_2O$	Trig	—	2.10	4–5	Fmwk	Zeolite
Chlorite^f	$(Mg,Fe,Mn,Al)_{12}[(Si,Al)_8O_{20}](OH)_{16}$	M	+/−	2.6–3.3	2–3	Sheet	Chlorite
Chloritoid	$(Fe,Mg,Mn)Al_2SiO_5(OH)_2$	M	+/−	3.46–3.80	6.5	Island	
Chondrodite	$(Mg,Fe)_5(SiO_4)_2(F,OH)_2$	M	+	3.16–3.26	6.5	Island	Humite
Chrysotile	$Mg_3Si_2O_5(OH)_4$	M/O	—	2.5	2.5	Sheet	Kaolinite–serpentine
Clinohumite	$(Mg,Fe)_9(SiO_4)_4(F,OH)_2$	M	+	3.17–3.35	6	Island	Humite
Clinozoisite	$Ca_7Al_3Si_3O_{12}(OH)$	M	+	3.21–3.38	6.5	Island	Epidote
Cordierite	$(Mg,Fe)_2Al_4Si_5O_{18}$.	O	−/+	2.53–2.78	7	Ring	
Cristobalite	SiO_2	Tet	—	2.33	6–7	Fmwk	Silica
Cummingtonite	$(Mg,Fe,Mn)_7Si_8O_{22}(OH)_2$	M	+	3.1–3.6	5–6	Chain	Amphibole
Diopside	$MgCaSi_2O_6$	M	+	3.22–3.38	5.5–6.5	Chain	Pyroxene
Eckermannite	$Na_3(Mg,Fe)_4AlSi_8O_{22}(OH)_2$	M	—	3.00–3.20	5–6	Chain	Amphibole
Edenite	$NaCa_2(Mg,Fe)_5Si_7AlO_{22}(OH)_2$	M	—	3.05–3.37	5–6	Chain	Amphibole
Enstatite	$Mg_2Si_2O_6$	O	+	3.21	5–6	Chain	Pyroxene

a CS = crystal system, b OS = optic sign, c D = density, d H = hardness, e Struc. = structure. (See notes at end of table.)

(*cont.*)

9

Mineral	Formula	CS^a	OS^b	D^c	H^d	Struc.^{e}	Group
Epidote	$Ca_2(Al,Fe)_3(SiO_4)_3OH$	M	−	3.38–3.49	6	Island	Epidote
Fayalite	Fe_2SiO_4	O	+	3.2	7	Island	Olivine
Ferrosilite	$Fe_2Si_2O_6$	O	−	3.96	5–6	Chain	Pyroxene
Forsterite	Mg_2SiO_4	O	−	4.4	6.5	Island	Olivine
Gedrite	$(Mg,Fe)_5Al_2Si_6Al_2O_{22}(OH)_2$	O	−/+	3.15–3.26	5.5–6	Chain	Amphibole
Gehlenite	$CaAl_2SiO_7$	Tet	−	3.07	5–6	Island	Melilite
Glauconite	$(K,Na)(Fe,Al,Mg)_2(Si,Al)_4O_{10}(OH)_2$	M	−	2.4–2.95	2	Sheet	Mica
Glaucophane	$Na_2(Mg,Fe)_3Al_2Si_8O_{22}(OH)_2$	M	−	3.05–3.15	6	Chain	Amphibole
Grossular	$Ca_3Al_2(SiO_4)_3$	C	−	3.954	6.5–7	Island	Garnet
Grunerite	$(Fe,Mg)_7Si_8O_{22}(OH)_2$	M	−	3.40–3.60	5–6	Chain	Amphibole
Hastingsite	$NaCa_2(Fe^{2+},Mg)_4Fe^{3+}(Si_6Al_2)O_{22}(OH)_2$	M	−	3.35–3.5	5–6	Chain	Amphibole
Haüyne	$(Na,Ca)_{4-8}Al_6Si_6(O,S)_{24}(SO_4,Cl)_{1-2}$	C	−	2.44–2.50	5.5–6	Fmwk	Sodalite
Hedenbergite	$CaFeSi_2O_6$	M	+	3.50–3.56	6	Chain	Pyroxene
Heulandite	$(Na,Ca)_{2-3}Al_3(Al,Si)_2Si_{13}O_{36}.12H_2O$	M	−/+	2.1–2.2	3.5–4	Fmwk	Zeolite
Hornblende^g	$Ca_2(Mg,Fe)_4AlSi_7AlO_{22}(OH)_2$	M	−	3.12–3.30	5–6	Chain	Amphibole
Humite	$(Mg,Fe)_7(SiO_4)_3(F,OH)_2$	O	+	3.20–3.32	6	Island	Humite
Hydrogrossular	$Ca_3Al_2(SiO_4)_{3-x}(OH)_{4x}$	C	−	3.13–3.59	6–7	Island	Garnet
Hypersthene^h	Synonym of enstatite or ferrosilite						
Illite^b	$K_yAl_4(Si_{8-y},Al_y)O_{20}(OH)_4$	M	−	2.6–2.9	1.2	Sheet	Mica

		CS[a]	OS[b]	D[c]	H[d]	Struc.[e]	
Jadeite	$Na(Al,Fe)Si_2O_6$	M	+	3.24–3.43	6	Chain	Pyroxene
Kaersutite	$NaCa_2(Mg,Fe)_4TiAl_2Si_6(O,OH)_{24}$	M	–	~3.3	5–6	Chain	Amphibole
Kaolinite	$Al_2Si_2O_5(OH)_4$	Tric	–	2.6–2.68	2–2.5	Sheet	Kaolinite–serpentine
Katophorite	$(Ca,Na,K)_3(Mg,Fe,Al)_5(Si,Al)8O_{22}(OH)_2$	M	–	3.20–3.50	5–6	Chain	Amphibole
Kyanite	Al_2SiO_5	Tric	–	3.53–3.65	5.5–7	Island	
Laumontite	$CaAl_2Si_4O_{12}.4H_2O$	M	–	2.27	4	Fmwk	Zeolite
Lawsonite	$CaAl_2Si_2O_7(OH)_2.H_2O$	O	+	3.05–3.12	6	Island	
Lazurite	$(Na,Ca)_8(AlSiO_4)_6(S,SO_4,Cl)_{1-2}$	C		2.4–2.45	5–5.5	Fmwk	Sodalite
Lepidolite	$K(Li,Al)_3(Si,Al)_4O_{10}(F,OH)_2$	M	–	2.8–2.9	2.5–4	Sheet	Mica
Leucite	$KAlSi_2O_6$	Tet	+	2.47	5.5–6	Fmwk	
Lizardite	$Mg_3Si_2O_5(OH)_4$	M	–	~2.5	2.5	Sheet	Kaolinite–serpentine
Marialite	$3NaAlSi_3O_8.NaCl$	Tet	–	2.50–2.62	5–6	Fmwk	Scapolite
Meionite	$3CaAl_2Si_2O_8.CaCO_3$	Tet	–	2.74–2.78	5–6	Fmwk	Scapolite
Melilite	$(Ca,Na)_2(Al,Mg)(Si,Al)_2O_7$	Tet	+/–	2.95–3.05	5–6	Island	Melilite
Mesolite	$Na_2Ca_2Al_6Si_9O_{30}.8H_2O$	O	+	2.26	5	Fmwk	Zeolite
Microcline	$KAlSi_3O_8$	Tric	–	2.54–2.57	6–6.5	Fmwk	Feldspar
Monticellite	$CaMgSiO_4$	O	–	3.05–3.27	5.5	Island	Olivine
Montmorillonite	$(Na,Ca)_{0.33}(Al,Mg)_2Si_4O_{10}(OH)_2.nH_2O$	M	–	2–3	1–2	Sheet	Smectite

a CS = crystal system, b OS = optic sign, c D = density, d H = hardness, e Struc. = structure. (See notes at end of table.)

(cont.)

Table 9.6 (*cont.*)

Mineral	Formula	CS^a	OS^b	D^c	H^d	Struc.^{e}	Group
Mullite	$Al_6Si_2O_{13}$	O	+	3.11–3.26	6–7	Island	
Muscovite	$KAl_2(Si_3Al)O_{10}(OH)_2$	M	–	2.77–2.88	2.5–3	Sheet	Mica
Natrolite	$Na_2Al_2Si_3O_{10}.2H_2O$	O	+	2.20–2.60	5–5.5	Fmwk	Zeolite
Nepheline	$(Na,K)AlSiO_4$	H	–	2.56–2.66	5.5–6	Fmwk	
Nosean	$Na_8Al_6Si_6O_{24}(SO_4)$	C		2.3–2.4	5.5	Fmwk	Sodalite
Olivine	$(Mg,Fe)_2SiO_4$	O	+/–	3.22–4.39	6.5–7	Island	Olivine
Omphacite	$(Ca,Na)(Mg,Fe,Al)Si_2O_6$	M	+	3.16–3.43	5–6	Chain	Pyroxene
Orthoclase	$KAlSi_3O_8$	M	–	2.57	6	Fmwk	Feldspar
Pargasite	$NaCa_2(Mg,Fe)_4Al(Si_6Al_2)O_{22}(OH)_2$	M	+	3.04–3.17	5–6	Chain	Amphibole
Pectolite	$NaCa_2Si_3O_8OH$	Tric	+	2.86–2.90	4.5–5	Chain	Pyroxene
Phlogopite	$KMg_3AlSi_3O_{10}(F,OH)_2$	M	–	2.76–2.90	2–2.5	Sheet	Mica
Piemontite	$Ca_2(Al,Fe)_3Si_3O_{12}OH$	M	+	3.38–3.61	6	Island	Epidote
Pigeonite	$(Mg,Fe,Ca)_2Si_2O_6$	M	+	3.17–3.46	6	Chain	Pyroxene
Prehnite	$Ca_2Al_2Si_3O_{10}(OH)_2$	O	+	2.9–3	6–6.5	Sheet	
Pumpellyite	$Ca_2FeAl_2SiO_4Si_2O_7(OH)_2.H_2O$	M	+/–	3.16–3.25	5–6	Island	
Pyrope	$Mg_3Al_2Si_3O_{12}$	C		3.58	7	Island	Garnet
Pyrophyllite	$Al_2Si_4O_{10}(OH)_2$	M/Tric	–	2.65–2.90	1–2	Sheet	Pyrophyllite–Talc
Pyroxmangite	$MnSiO_3$	Tric	+	3.61–3.80	5.5–6	Chain	
Quartz	SiO_2	Trig	+	2.65	7	Fmwk	Silica
Rhodonite	$MnSiO_3$	Tric	+	3.57–3.76	5.5–6.5	Chain	

6

Name	Formula	CS	OS	D	H	Struc.	
Richterite	$Na_2Ca(Mg,Fe)_5Si_8O_{22}(OH)_2$	M	–	2.97–3.45	5–6	Chain	Framework
Riebeckite	$Na_2(Fe^{2+},Mg)_3Fe^{3+}_2Si_8O_{22}(OH)_2$	M	+/–	3.15–3.50	5	Chain	Amphibole
Sanidine	$(K,Na)AlSi_3O_8$	M	–	2.56–2.62	6	Fmwk	Feldspar
Sapphirine	$(Mg,Al)_4(Al,Si)_3O_{10}$	M	–/+	3.40–3.58	7.5	Chain	
Scapolite	See marialite and meionite						
Scolecite	$CaAl_2Si_3O_{10}.3H_2O$	M	–	2.25–2.29		Fmwk	Zeolite
Sillimanite	Al_2SiO_5	O	+	3.23–3.27	6.5–7.5	Island	
Sodalite	$Na_4Al_3Si_3O_{12}Cl$	C		2.27–2.33	5.5–6	Fmwk	Sodalite
Spessartine	$Mn_3Al_2(SiO_4)_3$	C		4.19	7	Island	Garnet
Sphene	Synonym of titanite						
Spodumene	$LiAlSi_2O_6$	M	+	3.03–3.23	6.5–7	Chain	Pyroxene
Staurolite	$(Fe,Mg,Zn)_2Al_9(Si,Al)_4O_{22}(OH)_2$	M	+	3.74–3.83	7	Island	
Stilbite	$NaCa_2Al_5Si_{13}O_{36}.14H_2O$	M	–	2.1–2.2	3.5–4	Fmwk	Zeolite
Stilpnomelane	$K(Fe,Mg)_8(Si,Al)_{12}(O,OH)_{27}$	Tric	–	2.6–2.9	3–4	Sheet	
Talc	$Mg_3Si_4O_{10}(OH)_2$	Tric	–	2.58–2.83	1	Sheet	Pyrophyllite–talc
Tephroite	Mn_2SiO_4	O	–	3.7–4.1	6–6.5	Island	Olivine

a CS = crystal system, *b* OS = optic sign, *c* D = density, *d* H = hardness, *e* Struc. = structure. (See notes at end of table.)

(*cont.*)

9

Table 9.6 (*cont.*)

Mineral	Formula	CS^a	OS^b	D^c	H^d	Struc.^{e}	Group
Thomsonite	$NaCa_2Al_5Si_5O_{20}.6H_2O$	O	+	2.3	5	Fmwk	Zeolite
Titanite	$CaTiSiO_5$	M	+	3.5–3.6	5	Island	
Topaz	$Al_2SiO_4(F,OH)_2$	O	+	3.49–3.57	8	Island	
Tourmaline	$(Na,Ca)(Li,Mg,Fe^{2+},Al)_3(Al,Fe^{3+})_6$ $B_3Si_6O_{27}(O,OH,F)_4$	Trig	–	3.03–3.15	7	Ring	
Tremolite	$Ca_2(Mg,Fe)_5Si_8O_{22}(OH)_2$	M	–	2.99–3.48	5–6	Chain	Amphibole
Tridymite	SiO_2	O	+	2.26	7	Fmwk	Silica
Tschermakite	$Ca_2(Mg,Fe)_3Al_2(Si_6Al_2)O_{22}(OH)_2$	M	+/–	~3.15	5–6	Chain	Amphibole
Uvarovite	$Ca_3Cr_2(SiO_4)_3$	C		3.83	6.5–7	Island	Garnet
Vermiculite	$(Mg,Fe,Al)_3(Al,Si)_4O_{10}(OH)_2.4H_2O$	M	–	~2.3	~1.5	Sheet	
Vesuvianite	$Ca_{10}(Mg,Fe)_2Al_4Si_9O_{34}(OH)_4$	Tet	–	3.32–3.34	6–7	Island	
Wollastonite	$CaSiO_3$	Tric	–	2.86–3.09	4.5–5	Chain	Pyroxene
Zircon	$ZrSiO_4$	Tet	+	4.6–4.7	7.5	Island	
Zoisite	$Ca_2Al_3(SiO_4)_3OH$	O	+	3.15–3.37	6–7	Island	Epidote

Notes: *a* CS = crystal system: C = cubic, Tet = tetragonal, O = Orthorhombic, M = Monoclinic, H = Hexagonal, Trig = Trigonal, Tric = Triclinic. *b* OS = optic sign. *c* D = density. *d* H = hardness. *e* Struc. = structure – see Table 8.5; Fmwk = framework structure. *f* Chlorite is strictly a name for a group of minerals including clinochlore and chamosite. *g* The name hornblende is also applied to a wider compositional range – either with greater magnesium or iron. *h* Illite is a series name for a range of compositions in which *y* is usually between 1 and 1.5.

Source: See Table 9.7.

6

Table 9.7 Common non-silicate minerals

Mineral	Formula	Crystal system[a]	Optic sign	Density	Hardness	Group
Anatase	TiO_2	Tet	−	3.82–3.97	5.5–6	
Anhydrite	$CaSO_4$	O	+	2.9	3–3.5	
Ankerite	$Ca(Fe,Mg,Mn)(CO_3)_2$	Trig	−	2.93–3.10	3.5–4	Dolomite
Apatite	$Ca_5(PO_4)_3(OH,F,Cl)$	H	−	3.1–3.4	5	Apatite
Aragonite	$CaCO_3$	O	−	2.95	3.5–4	Aragonite
Azurite	$Cu_3(CO_3)_2(OH)_2$	M	+	3.8–3.9	3.5–4	
Baryte	$BaSO_4$	O	+	4.5	2.5–3.5	Baryte
Böhmite	$\gamma\text{-AlO(OH)}$	O	+	3	3.5–4	
Brookite	TiO_2	O	+	4.1–4.2	5.5–6	
Brucite	$Mg(OH)_2$	Trig	+	2.39	2.5	
Calcite	$CaCO_3$	Trig	−	2.7	3	Calcite
Cassiterite	SnO_2	Tet	+	6.98–7.02	6–7	
Celestine	$SrSO_4$	O	+	4	3–3.5	Baryte
Chalcopyrite	$CuFeS_2$	Tet		4.1–4.3	3.5–4.5	
Chromite	$FeCr_2O_4$	C		5	5.5	Spinel
Corundum	$\alpha\text{-Al}_2O_3$	Trig	−	3.98–4.02	9	Hematite
Diaspore	$\alpha\text{-AlO(OH)}$	O	+	3.3–3.5	6.5–7	
Dolomite	$CaMg(CO_3)_2$	Trig	−	2.9	3.5	Dolomite

(cont.)

Table 9.7 (*cont.*)

Mineral	Formula	Crystal system[a]	Optic sign	Density	Hardness	Group
Fluorite	CaF_2	C		3.2	4	
Franklinite	$(Zn,Mn^{2+},Fe^{2+})(Fe^{3+},Mn^{3+})_2O_4$	C		5.34	~6	Spinel
Galena	PbS	C		7.5–7.6	2.5	
Gibbsite	$Al(OH)_3$	M	+	2.4	2.5–3.5	
Goethite	α-$FeO(OH)$	O	−	4.3	5–5.5	
Gypsum	$CaSO_4.2H_2O$	M	+	2.30–2.37	2	
Halite	$NaCl$	C		2.2	2.5	
Hematite	α-Fe_2O_3	Trig	−	5.25	5–6	Hematite
Hercynite	$Fe^{2+}Al_2O_4$	C		4.40	7.5–8	Spinel
Huntite	$Mg_3Ca(CO_3)_4$	Trig	−	2.7		
Ilmenite	$FeTiO_3$	Trig	−	4.70–4.79	5–6	
Lepidocrocite	γ-$FeO(OH)$	O	−	4.09	5	
Magnesite	$MgCO_3$	Trig	−	3.0	4	Calcite
Magnetite	$Fe^{2+}Fe_2^{3+}O_4$	C		5.2	5.5–6.5	Spinel
Malachite	$Cu_2CO_3(OH)_2$	M	−	3.7–4.1	3.5–4	
Monazite	$(Ce,La,Nd,Th)PO_4$	M	+	5–5.4	5	
Periclase	MgO	C		3.56–3.68	5.5–6	Periclase
Perovskite	$(Ca,Na,Fe^{2+},Ce,Sr)(Ti,Nb)O_3$	O	+	4–4.3	5.5	Perovskite
Pyrite	FeS_2	C		4.95–5.03	6–6.5	Pyrite

6

Pyrrhotite	FeS	M		4.6	3.5–4.5	
Rhodocrosite	MnCO$_3$	Trig	–	3.7	3.5–4	Calcite
Rutile	TiO$_2$	Tet	+	4.23	6–6.5	Rutile
Siderite	FeCO$_3$	Trig	–	4.0	4–4.5	Calcite
Smithsonite	ZnCO$_3$	Trig	–	4.43	4–4.5	Calcite
Sphalerite	(Zn,Fe)S	C		4.1	3.5–4	
Spinel	MgAl$_2$O$_4$	C		3.55	7.5–8	Spinel
Strontianite	SrCO$_3$	O	–	3.72	3.5	Aragonite
Troilite	FeS	H		4.6	3.5–4.5	
Witherite	BaCO$_3$	O	–	4.3	3.5	Aragonite

Note: *a* C = cubic, Tet = tetragonal, O = Orthorhombic, M = Monoclinic, H = Hexagonal, Trig = Trigonal.

Sources: Anthony, J. W. et al. (1990). *Handbook of Mineralogy*, vol. 1. *Elements, Sulfides, Sulfosalts*. Tuscon, AZ: Mineral Data Publishing. Anthony, J. W. et al. (1995). *Handbook of Mineralogy*, vol. II, *Silicates*. Anthony, J. W. et al. (1997). *Handbook of Mineralogy*, Vol. III, *Halides, Hydroxides, Oxides*. Clark, A. M. (1993). *Hey's Mineral Index*, 3rd edn. London: Chapman and Hall. Deer, A. et al. (1992). *An Introduction to the Rock-Forming Minerals*, 2nd edn. Harlow: Longman Group. Chang, L. L. Y. et al. (1996). *Rock-Forming Minerals*, vol. 5B, *Non-silicates: Sulphates, Carbonates, Phosphates, Halides*. Harlow: Longman Group. Klein, C. (2002). *Manual of Mineral Science*, 22nd edn. New York: John Wiley. Mandarino, J. M. and Back, M. E. (2004). *Fleischer's Glossary of Mineral Species*, 2004, 9th edn. Tuscon, AZ: The Mineralogical Record Inc.

9

Table 9.8 Moh's scale of mineral hardness

1 to 10 increasing hardness

1 Talc	6 Orthoclase feldspar
2 Gypsum	7 Quartz
3 Calcite	8 Topaz
4 Fluorite	9 Corundum
5 Apatite	10 Diamond

Table 9.9 Magnetic properties of common magnetic minerals

Mineral	Composition	Magnetic order	T_C^a (°C)	M_s^b (10^3 A/m)
Oxides				
Magnetite	Fe_3O_4	Ferrimagnetic	575–585	480
Titanomagnetite (TM60)	$Fe_{2.4}Ti_{0.6}O_4$	Ferrimagnetic	150	125
Ulvöspinel	Fe_2TiO_2	Anti-ferromagnetic	−153	
Hematite	αFe_2O_3	Canted anti-ferromagnetic	675	~2.5
Ilmenite	$FeTiO_2$	Anti-ferromagnetic	−233	
Maghemite	γFe_2O_3	Ferrimagnetic	~590–675	380
Sulphides				
Pyrrhotite	Fe_7S_8	Ferrimagnetic	320	~80
Greigite	Fe_3S_4	Ferrimagnetic	~333	125
Oxyhydroxides				
Goethite	$\alpha FeOOH$	Anti-ferromagnetic – weak FM	~120	~2
Lepidocrocite	$\gamma FeOOH$	Anti-ferromagnetic ?	−196	
Metals & Alloys				
Iron	Fe	Ferromagnetic	770	1715
Nickel	Ni	Ferromagnetic	358	484
Cobalt	Co	Ferromagnetic	1131	1422

Notes: a T_C = Curie or Néel temperature. b M_s = saturation magnetisation at room temperature.

Source: This table was produced by Conall MacNiocaill of Oxford University, UK (2007).

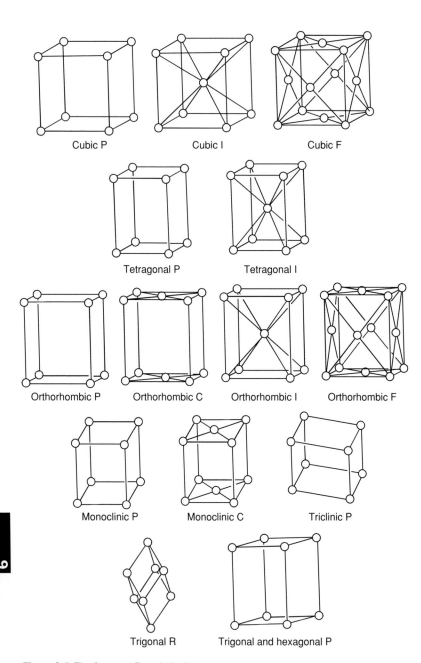

Figure 9.1 The fourteen Bravais lattices.

10 Resources

Table 2.9 on the classification and properties of coal can be found in Chapter 2. Mineral data can be found in Chapter 9.

10

Table 10.1 Element and mineral resources

Commodity	Annual production[a]	Reserve base[b]	Top two producing countries
	(thousand tonnes)		
Antimony	117	3900	China, South Africa
Asbestos	2.2		Russia, China
Barite	7620	740 000	China, India
Bauxite	165 000	32 000 000	Australia, Brazil
Beryllium	0.114		USA, China
Bismuth	5.2	680	China, Mexico
Bromine	580	Large	USA, Israel
Chromium	18 000	Large	South Africa, Kazakhstan
Clays			
Bentonite	10 700	Large	USA, Greece
Fuller's Earth	5400	Large	USA, Spain
Kaolin	44 500	Large	USA, CIS[c]
Cobalt	0.0524	13 000	Congo, Zambia
Copper	14 900	940 000	Chile, USA
Diamond[d]	74[e]	1250[e]	Australia, Russia
Diatomite	1950	Large	USA, China
Feldspar	11 500		Italy, Turkey
Fluorspar	5080	480 000	China, Mexico
Garnet	312	Large	Australia, India
Gold	2.450	90	South Africa, Australia
Graphite	992	290 000	China, India
Gypsum	110 000	Large	USA, Iran
Helium	166[f]	39 000[f]	USA, Russia
Iodine	26.2	27 000	Chile, Japan
Iron ore	1 520 000	370 000 000	China, Brazil
Kyanite[g]	350	Large	South Africa, USA
Lead	3280	140 000	China, Australia
Lithium	20.4	11 000	Chile, Australia
Manganese	9790	430 000	South Africa, Australia
Mercury	1.1	240	China, Kyrgyzstan
Mica (flakes)	290	Large	Russia, USA
(sheet)	5.2	Large	India, Russia

10

Table 10.1 (*cont.*)

Commodity	Annual production[a]	Reserve base[b]	Top two producing countries
	(thousand tonnes)		
Molybdenum	163	19 000	USA, Chile
Nickel	1500	140 000	Russia, USA
Niobium	33.9	5200	Brazil, Canada
Phosphate rock	148 000	50 000 000	USA, Morocco–W. Sahara
Pt-group metals	218	80 000	South Africa, Russia
Potash	31 000	17 000 000	Canada, Russia
Pumice	15 900	Large	Italy, Greece
Rare earths	105	150 000	China, India
Rhenium	0.043	10	Chile, Kazakhstan
Selenium	1.35	170	Japan, Canada
Silicon	5100	Large	China, Russia
Silver	20.3	570	Peru, China
Strontium	520	12 000	Spain, Mexico
Sulphur	64 000	Large	USA, Canada
Talc	8360	Large	China, Republic of Korea
Tantalum	1.91	Adequate	Australia, Mozambique
Tin	280	11 000	China, Indonesia
Titanium[b] (mineral concentrate)			
Ilmenite	4800	1 200 000	Australia, South Africa
Rutile	360	100 000	Australia, South Africa
Tungsten	76.5	6200	China, Russia
Vanadium	42.5	38 000	South Africa, China
Vermiculite	520	Large	South Africa, USA
Yttrium	2.4	610	China, India
Zinc	10 100	460 000	China, Australia
Zirconium	870	72 000	Australia, South Africa

Notes: *a* Estimated for the year 2005. *b* Reserve base: that part of an identified resource that meets the specified minimum physical and chemical criteria related to current mining and production practices, including those for grade, quality, thickness and depth. The reserve base is the in-place demonstrated (measured plus indicated) resource from which reserves are estimated. It includes those resources that are currently economic, marginally economic and some of those that are currently subeconomic. *c* Commonwealth of Independent States. *d* Industrial grade diamond. *e* Units of million carats (1 carat = 200 mg) *f* Units of million cubic metres. *g* Includes related aluminium silicates such as andalusite. *h* As TiO_2.

Sources: US Geological Survey web site http://minerals.usgs.gov/minerals as at September 2006.

Table 10.2 Oil reserves and production (for year 2007)

Countries		Reserves[a]		Production[b]
		Tonnes $\times 10^9$	Barrels $\times 10^9$	Barrels per day $\times 10^6$
N. America				
US		3.6	29.4	6.879
Canada		4.2	27.7	3.309
Mexico		1.7	12.2	3.477
	Total	9.5	69.3	13.665
S. and Cent. America				
Argentina		0.4	2.6	0.698
Brazil		1.7	12.6	1.833
Ecuador		0.6	4.3	0.520
Venezuela		12.5	87.0	2.613
Other		0.6	4.7	0.970
	Total	15.9	111.2	6.633
Europe and Eurasia				
Azerbaijan		1.0	7.0	0.868
Kazakhstan		5.3	39.8	1.490
Norway		1.0	8.2	2.556
Russian Federation		10.9	79.4	9.978
United Kingdom		0.5	3.6	1.636
Other		0.8	5.7	1.307
	Total	19.4	143.7	17.835
Middle East				
Iran		19.0	138.4	4.401
Iraq		15.5	115.0	2.145
Kuwait		14.0	101.5	2.626
Oman		0.8	5.6	0.718
Qatar		3.6	27.4	1.197
Saudi Arabia		36.3	264.2	10.413
Syria		0.3	2.5	0.394
United Arab Emirates		13.0	97.8	2.915

Table 10.2 (*cont.*)

Countries		Reserves[a]		Production[b]
		Tonnes $\times 10^9$	Barrels $\times 10^9$	Barrels per day $\times 10^6$
Yemen		0.4	2.8	0.336
Other			0.1	0.032
	Total	102.9	755.3	25.176
Africa				
Algeria		1.5	12.3	2.000
Angola		1.2	9.0	1.723
Rep. of Congo (Brazzaville)		0.3	1.9	0.222
Egypt		0.5	4.1	0.710
Gabon		0.3	2.0	0.230
Libya		5.4	41.5	1.848
Nigeria		4.9	36.2	2.356
Sudan		0.9	6.6	0.457
Other		0.5	3.9	0.772
	Total	15.6	117.5	10.318
Asia Pacific				
Australia		0.4	4.2	0.561
China		2.1	15.5	3.743
India		0.7	5.5	0.801
Indonesia		0.6	4.4	0.969
Malaysia		0.7	5.4	0.755
Vietnam		0.5	3.4	0.340
Other		0.4	2.6	0.737
	Total	5.4	40.8	7.907
	World total	168.6	1237.9	81.533

Notes: *a* Proved oil reserves are those estimated, with reasonable certainty, which can be recovered in the future from known reserves under existing economic and geological conditions. They include crude oil and natural gas liquids. *b* Production figures include crude oil, shale oil and natural gas liquids. The last contributed ~11% to the World's total oil production. Totals allow for figure rounding.

Source: BP p.l.c. – *Statistical Review of World Energy 2008*, with permission. BP usually publish their annual statistical review in June each year. The latest version can usually be found via their website www.bp.com

Table 10.3 Coal reserves and production (for year end 2007)

Country	Reserves[a] (million tonnes)	Production (million tonnes)	Country	Reserves[a] (million tonnes)	Production (million tonnes)
North America			*Africa and Middle East*		
US	242 721	1039.2	South Africa	48 000	269.4
Canada	6 578	69.4	Other Africa	1 605	3.8
Mexico	1 211	12.2	Middle East	1 386	0.8
Total	250 510	1120.8	Total	50 991	274.0
South and Central America			*Europe and Eurasia*		
Brazil	7 068	5.9	Bulgaria	1 996	30.4
Colombia	6 959	71.7	Czech Republic	4 501	62.6
Venezuela	479	8.0	Germany	6 708	201.9
Other	1 770	1.0	Greece	3 900	62.5
Total	16 276	86.6	Hungary	3 302	9.8
Asia Pacific			Kazakhstan	31 300	94.4
Australia	76 600	393.9	Poland	7 502	145.8
China	114 500	2536.7	Russian Federation	157 010	314.2
India	56 498	478.2	Turkey	1 814	76.6
Indonesia	4 328	174.8	Ukraine	33 873	76.3
Pakistan	1 982	3.6	United Kingdom	155	17
Thailand	1 354	18.3			
Vietnam	150	41.2	Other	20 185	122.7
Other	2 052	53.2	Total	272 246	1214.2
Total	257 464	3699.9	**World totals**	**847 488**	**6395.6**

Note: *a* Proved reserves of coal are those quantities that geological and engineering information indicates with reasonable certainty can be recovered in the future from known deposits under existing economic and operating conditions. The table includes anthracite, bituminous, sub-bituminous and lignite deposits.

Source: See Table 10.2.

10

Table 10.4 Gas reserves and production (for end year 2007)

Country	Reserves[a] m³ × 10¹²	Production m³ × 10⁹	Country	Reserves[a] m³ × 10¹²	Production m³ × 10⁹
North America			Turkmenistan	2.67	67.4
USA	5.98	545.9	Ukraine	1.03	19
Canada	1.63	183.7	United Kingdom	0.41	72.4
Mexico	0.37	46.2			
Total	7.98	775.8	Uzbekistan	1.74	58.5
			Other	0.43	11
South and Central America			Total	59.41	1075.7
Argentina	0.44	44.8			
Bolivia	0.74	13.5	*Middle East*		
Brazil	0.36	11.3	Bahrain	0.09	11.5
Colombia	0.13	7.7	Iran	27.80	111.9
Peru	0.36		Iraq	3.17	
Trinidad & Tobago	0.48	39.0	Kuwait	1.78	12.6
			Oman	0.69	24.1
Venezuela	5.15	28.5	Qatar	25.60	59.8
Other	0.07	6.1	Saudi Arabia	7.17	75.9
Total	7.73	150.8	Syria	0.29	5.3
			United Arab Emirates	6.09	49.2
Europe and Eurasia					
Azerbaijan	1.28	10.3	Yemen	0.49	
Denmark	0.12	9.2	Other	0.05	5.5
Germany	0.14	14.3	Total	73.21	355.8
Italy	0.09	8.9			
Kazakhstan	1.90	27.3	*Africa*		
Netherlands	1.25	64.5	Algeria	4.52	83.0
Norway	2.96	89.7	Egypt	2.06	46.5
Poland	0.11	4.3	Libya	1.50	15.2
Romania	0.63	11.6	Nigeria	5.30	35.0
Russian Federation	44.65	607.4	Other	1.21	10.7
			Total	14.59	190.4

(cont.)

10

Table 10.4 (*cont.*)

Country	Reserves[a] $m^3 \times 10^{12}$	Production $m^3 \times 10^9$	Country	Reserves[a] $m^3 \times 10^{12}$	Production $m^3 \times 10^9$
Asia Pacific					
Australia	2.51	40.0	Myanmar	0.60	14.7
Bangladesh	0.39	16.3	Pakistan	0.85	30.8
Brunei	0.34	12.3	Papua New Guinea	0.44	
China	1.88	69.3			
India	1.06	30.2	Thailand	0.33	25.9
Indonesia	3.00	66.7	Vietnam	0.22	7.7
Malaysia	2.48	60.5	Other	0.37	17.1
			Total	14.46	391.5
World totals	**177.36**	**2940.0**			

Note: *a* Proved reserves are those quantitites that geological and engineering information indicates with reasonable certainty can be recovered in the future from known reserves under existing economic and operating conditions. Totals allow for figure rounding.

Source: As for Table 10.2

Table 10.5 Uranium resources and production

Country	Resources RAR^a, tU	IR^b, tU	Total, tU	Production tU/a
Algeria	19 500	na	19 500	na
Argentina	7 080	8 560	15 640	0
Australia	747 000	396 000	1 143 000	8 982
Brazil	157 700	121 000	278 700	300
Bulgaria	5 870	6 300	12 170	0
Canada	345 200	98 600	443 800	11 597
Central African Republic	12 000	na	12 000	na
China	38 019	21 704	59 723	730
Denmark	20 250	12 000	32 250	na
Finland	1 125	11 740	12 865	0
India	42 568	22 272	64 840	230
Jordan	30 375	48 600	78 975	na
Kazakhstan	513 897	302 202	816 099	3 719
Mongolia	46 200	15 750	61 950	0
Namibia	182 556	99 803	282 359	3 039
Niger	180 446	44 993	225 439	3 245
Russian Federation	131 750	40 652	172 402	3 280
South Africa	255 593	85 003	340 596	747
Spain	4 925	6 380	11 305	0
Sweden	4 000	6 000	10 000	0
Ukraine	66 706	23 130	89 836	800
USA	342 000	na	342 000	878
Uzbekistan	76 936	38 590	115 526	2 087
Other Countries	64 973	36 895	101 868	629
	3 296 669	1 446 174	4 742 843	40 263

Notes: *a* RAR = Reasonably Assured Resources – known deposits of delineated size, grade and configuration such that the quantities given could be recovered at a cost of <US$130/kg U with currently proven mining and processing technology. *b* IR = Inferred Resources – inferred to occur based on direct geological evidence, in extension of well-explored deposits, but where data are insufficient to allow them to be classified as RAR. Individual data are given for those countries where total reserves = 10 000 tU or greater. Reserve estimates are mostly as at January 2005, production estimates are for the year 2004. na = data not available.

Source: *Uranium 2005: Resources, Production and Demand.* (2006). Paris: A joint report by the OECD Nuclear Energy Agency and the International Atomic Energy Agency.

10

Table 10.6 Freshwater resources and use, by country

Country	Supply km³/year	Use	Country	Supply km³/year	Use
		Africa			
Algeria	14.3	6.07	Lesotho	5.2	0.05
Angola	184	0.35	Liberia	232.0	0.11
Benin	25.8	0.13	Libya	0.6	4.27
Botswana	14.7	0.19	Madagascar	337.0	14.96
Burkina Faso	17.5	0.80	Malawi	17.3	1.01
Burundi	3.6	0.29	Mali	100.0	6.55
Cameroon	285.5	0.99	Mauritania	11.4	1.70
Cape Verde	0.3	0.02	Mauritius	2.2	0.61
Central African Republic	144.4	0.03	Morocco	29.0	12.60
			Mozambique	216.0	0.63
Chad	43	0.23	Namibia	45.5	0.3
Comoros	1.2	0.01	Niger	33.7	2.18
Congo	832	0.03	Nigeria	286.2	8.01
Congo, Dem. Republic	1283	0.36	Rwanda	5.0	0.15
			Senegal	39.4	2.22
Cote d'Ivoire	81	0.93	Sierra Leone	160.0	0.38
Djibouti	0.3	0.02	Somalia	15.7	3.29
Egypt	86.8	68.30	South Africa	50.0	12.50
Equatorial Guinea	26	0.11	Sudan	154.0	37.32
			Swaziland	4.5	1.04
Eritrea	6.3	0.30	Tanzania	91.0	5.18
Ethiopia	110.0	5.56	Togo	14.7	0.17
Gabon	164.0	0.12	Tunisia	4.6	2.64
Gambia	8.0	0.03	Uganda	66.0	0.30
Ghana	53.2	0.98	Zambia	105.2	1.74
Guinea	226.0	1.51	Zimbabwe	20.0	4.21
Guinea–Bissau	31.0	0.18			
Kenya	30.2	1.58			

Table 10.6 (*cont.*)

Country	Supply km³/year	Use	Country	Supply km³/year	Use
		North and Central America			
Antigua and Barbuda	0.1	0.005	Haiti	14.0	0.99
			Honduras	95.9	0.86
Barbados	0.1	0.09	Jamaica	9.4	0.41
Belize	18.6	0.15	Mexico	457.2	78.22
Canada	3300.0	44.72	Nicaragua	196.7	1.30
Costa Rica	112.4	2.68	Panama	148.0	0.82
Cuba	38.1	8.20	St. Kitts and Nevis	0.02	na
Dominica	na	0.02			
Dominican Republic	21.0	3.39	Trinidad and Tobago	3.8	0.31
El Salvador	25.2	1.28	USA	3069.0	477.00
Guatemala	111.3	2.01			
		South America			
Argentina	814.0	29.19	Guyana	241.0	1.64
Bolivia	622.5	1.44	Paraguay	336.0	0.49
Brazil	8233.0	59.30	Peru	1913.0	20.13
Chile	922.0	12.55	Suriname	122.0	0.67
Colombia	2132.0	10.71	Uruguay	139.0	3.15
Ecuador	432.0	16.98	Venezuela	1233.2	8.37
		Asia			
Afghanistan	65.0	23.26	Georgia	63.3	3.61
Armenia	10.5	2.95	India	1907.8	645.84
Azerbaijan	30.3	17.25	Indonesia	2838.0	82.78
Bahrain	0.1	0.30	Iran	137.5	72.88
Bangladesh	1210.6	79.40	Iraq	96.4	42.70
Bhutan	95.0	0.43	Israel	1.7	2.05
Brunei	8.5	0.09	Japan	430.0	88.43
Cambodia	476.1	4.08	Jordan	0.9	1.01
China	2829.6	549.76	Kazakhstan	109.6	35.00

(*cont.*)

10

Table 10.6 (*cont.*)

Country	Supply km³/year	Use	Country	Supply km³/year	Use
Korea, DPR	77.1	9.02	Russian Federation	4498.0	76.68
Korea, Republic	69.7	18.59	Saudi Arabia	2.4	17.32
Kuwait	0.02	0.44	Singapore	0.6	0.19
Kyrgyz Republic	46.5	10.08	Sri Lanka	50.0	12.61
Laos	333.6	3.00	Syria	46.1	19.95
Lebanon	4.8	1.38	Taiwan	67.0	na
Malaysia	580.0	9.02	Tajikistan	99.7	11.96
Maldives	0.03	0.003	Thailand	409.9	82.75
Mongolia	34.8	0.44	Turkey	234.0	39.78
Myanmar	1045.6	33.23	Turkmenistan	60.9	24.65
Nepal	210.2	10.18	United Arab Emirates	0.2	2.30
Oman	1.0	1.36	Uzbekistan	72.2	58.34
Pakistan	233.8	169.39	Vietnam	891.2	71.39
Philippines	479.0	28.52	Yemen	4.1	6.63
Qatar	0.1	0.29			

Europe

Country	Supply km³/year	Use	Country	Supply km³/year	Use
Albania	41.7	1.71	France	189.0	33.16
Austria	84.0	3.67	Germany	188.0	38.01
Belarus	58.0	2.79	Greece	72.0	8.70
Belgium	20.8	7.44	Hungary	120.0	21.03
Bosnia and Herzegovina	37.5	na	Iceland	170.0	0.17
Bulgaria	19.4	6.92	Ireland	46.8	1.18
Croatia	105.5	na	Italy	175.0	41.98
Cyprus	0.4	0.21	Latvia	49.9	0.25
Czech Republic	16.0	1.91	Lithuania	24.5	3.33
Denmark	6.1	0.67	Luxembourg	1.6	0.06
Estonia	21.1	1.41	Macedonia	6.4	2.27
Finland	110.0	2.33	Malta	0.07	0.02
			Moldova	11.7	2.31

10

Table 10.6 (*cont.*)

Country	Supply km^3/year	Use	Country	Supply km^3/year	Use
		Europe (*cont.*)			
Netherlands	89.7	8.86	Slovenia	32.1	0.90
Norway	381.4	2.40	Spain	111.1	37.22
Poland	63.1	11.73	Sweden	179.0	2.68
Portugal	73.6	11.09	Switzerland	53.3	2.52
Romania	42.3^a	6.50	Ukraine	139.5	37.53
Serbia–Montenegro	208.5	na	United Kingdom	160.6	11.75
Slovakia	80.3^a	1.04			
		Oceania			
Australia	398.0	24.06	Papua New Guinea	801.0	0.1
Fiji	28.6	0.07			
New Zealand	397.0	2.11	Solomon Islands	44.7	na

Notes: Data are estimates made for the different countries from 1975 to 2003 with most being made in the years 2000–2003. They are for renewable surface water and groundwater supplies including surface inflows from neighbouring countries. The data are averages – actual amounts will vary each year. na = not available. *a* These estimates were published by EUROSTAT (2003). The estimates given for 'use' should only be taken as a guide.

Source: Gleick, P. H. (2006). *The World's Water 2006–2007. The Biennial Report on Freshwater Resources*. Washington, DC: Island Press, with permision.

10

11 Hazards

Tables on past major volcanic events and earthquakes can be found in Chapter 7.

Table 11.1 Volcanic Explosivity Index (VEI) and frequency

Index number	Descriptors		Volume of tephra (m^3)	Eruption cloud column height (km)	Average frequency	Example
	Scale	Nature				
0	Non-explosive	Effusive	$<10^4$	<0.1	Numerous	Mauna Loa, 1984
1	Small	Gentle	10^4–10^6	0.1–1	100 per year	Nyiragongo, 2002
2	Moderate	Explosive	10^6–10^7	1–5	15 per year	Tristan da Cunha, 1961
3	Moderate–large	Severe	10^7–10^8	3–15	2–3 per year	Surtsey, Iceland, 1963–67
4	Large	Violent	10^8–10^9	10–25	1 per two years	Mt Pelée, 1902
5	Very large	Cataclysmic	10^9–10^{10}	>25	1 per 10 years	Mt St. Helens 1980
6	Huge	Paroxysmal	10^{10}–10^{11}	>25	1 per 40 years	Krakatoa, 1883
7	Colossal	Paroxysmal	10^{11}–10^{12}	>25	1 per 200 years	Mt Tambora, 1815
8	Colossal		$>10^{12}$	>25		Yellowstone, ~640 000 BC

Notes: In general, the frequency of occurrence of a given class of eruptive magnitude is inversely proportional to the volcanic energy released by the eruptions. For a discussion on analysis of eruption frequency for the case of a specific volcano and its use in surveillance, see: De la Cruz-Ryna, S. (1996). Long-term probabilistic analysis of future explosive eruptions. In *Monitoring and Mitigation of Volcanic Hazards*, ed. R. Scarpa and R. I. Tilling. Berlin: Springer-Verlag, pp. 599–629.

Sources: modified from Decker, R. and Decker, B. (1998). *Volcanoes*, 3rd edn. New York: W. H. Freeman & Co. Newhall, C. G. and Self, S. (1982). *Journal of Geophysical Research*, **87**, 1231–1238. Pyle, D. M. (2000). Sizes of volcanic eruptions. In *Encyclopedia of Volcanoes*, ed. H. Sigurdsson. San Diego, CA: Academic Press, pp. 263–269.

11

Table 11.2 Deaths by volcanic hazards

	Deaths (%)	
Hazard	1600–1986	1900–1986
Lava flows	0.17	0.10
Tephra falls	1.64	4.49
Pyroclastic flows	10.72	47.18
Lahars	6.07	38.86
Seismic action	0.01	0.04
Tsunami	7.79	0.17
Atmospheric hazards		0.0003
Gas and acid rain	0.34	2.55
Indirect	70.07	4.22
Totals	96.81	97.61

Notes: The number of fatalities is dependent on many parameters, including density of population. For example, the Mt Pelée eruption (1902) caused ~30 000 deaths, while the Mt St. Helens (1980) eruption caused 57 deaths, despite the latter having a higher VEI value than the former.

Source: Blong, R. J. (1996). Volcanic hazards risk assessment. In *Monitoring and Mitigation of Volcano Hazards*, ed. R. Scarpa and R. I. Tilling. Berlin: Springer-Verlag, pp. 675–698.

Table 11.3 Earthquake frequency

Averages		Major occurrences 1973–2007	
Magnitude	Average annual occurrence	Magnitude	Number
≥8.0	1	≥8.5	5
7–7.9	17	8.0–8.49	33
6–6.9	134	7.5–7.99	142
5–5.9	1319	7.0–7.49	350
4–4.9	13 000[a]	6.5–6.99	1045
3–3.9	130 000[a]	6.0–6.49	3176
2–2.9	1 300 000[a]		

Note: Data are based on earthquakes from 1900. See Table 7.8 for major earthquakes. a Data for these magnitudes are estimated.

Source: USGS Earthquake Hazard Programme, accessed Dec. 2007. http://neic.usgs.gov/neis/epic

Table 11.4 Earthquakes: the European Macroseismic Scale (EMS98, short form)[a]

Intensity	Definition	Description of typical observed effects
I	Not felt	Not felt.
II	Scarcely felt	Felt only by a few individual people at rest in houses.
III	Weak	Felt indoors by a few people. People at rest feel a swaying or light trembling.
IV	Largely observed	Felt indoors by many people, outdoors by a very few. A few people are awakened. Windows, doors and dishes rattle.
V	Strong	Felt indoors by most, outdoors by a few. Many sleeping people awake. A few are frightened. Buildings tremble throughout. Hanging objects swing considerably. Small objects are shifted. Doors and windows swing open or shut.
VI	Slightly damaging	Many people are frightened and run outdoors. Some objects fall. Many houses suffer slight non-structural damage such as hair-line cracks and the fall of small pieces of plaster.
VII	Damaging	Most people are frightened and run outdoors. Furniture is shifted and objects fall from shelves in large numbers. Many well-built ordinary buildings suffer moderate damage: small cracks in walls, fall of plaster, parts of chimneys fall down; older buildings may show large cracks in walls and failure of fill-in walls.
VIII	Heavily damaging	Many people find it difficult to stand. Many houses have large cracks in walls. A few well-built ordinary buildings show serious failure of walls, while weak older structures may collapse.
IX	Destructive	General panic. Many weak constructions collapse. Even well-built ordinary buildings show very heavy damage: serious failure of walls and partial structural failure.

11

Table 11.4 (*cont.*)

Intensity	Definition	Description of typical observed effects
X	Very destructive	Many ordinary well-built buildings collapse.
XI	Devastating	Most ordinary well-built buildings collapse, even some with good earthquake resistant design are destroyed.
XII	Completely destructive	Almost all buildings are destroyed.

Note: *a* The European Seismological Commission of the Council of Europe has devised a semi-quantitative scale for assessing earthquake intensity. This is increasingly being used internationally when assessing hazards. The above is the 1998 revision and is the official 'short form'. The complete form can be found in the source given below. The scale is similar to the Mercalli scale that it now seems to be replacing in usage. Quantitative scales used in geophysical and related studies can be found in Chapter 3.

Source: http://www.gfz-potsdam.de/pb5/pb53/projekt/ems/index_d.html Accessed 3/2008.

Table 11.5 Tsunami magnitude and occurrence

Magnitude, M^a	Number of events[b]	%
−5	1	0.1
−3	67	8.4
−2	63	7.9
−1	113	14.2
0	43	5.4
1	234	29.4
2	135	17.0
3	67	8.4
4	46	5.8
5	19	2.4
6	7	0.9
Total	795	99.9

Notes: *a* The magnitude given here is that by Iida *et al.*, defined as $M = \log_2 H$ where H is the maximum run-up height (metres) of the wave. Other scales exist but some, such as that by Soloviev, are not used consistently. For a discussion on magnitude and intensity scales see Bryant, E. (2001). *Tsunami. The Underrated Hazard.* Cambridge: Cambridge University Press *b* The number of events is based on data from the source below.

Source: The US National Geophysical Data Center (NGDC), Historical Tsunami Database (accessed Jan. 2008), http://www.ngdc.noaa.gov/hazard/tsu_db.shtml This incorporates US Government material that is not subject to copyright protection.

11

Table 11.6 Significant tsunamis since 1940

Date	Location	Latitude	Longitude	Cause[a]	Earthquake magnitude	Max. water height[b](m)	Tsunami magnitude[c]	Deaths
1941, June	India, Andaman Sea	12.5N	92.5E	Earthqk	7.6	na	na	5 000
1944, Dec.	Japan, Kii Peninsula	34.0N	137.1E	Earthqk	8.1	10.0	2.9	1 223
1946, Aug.	Dominican Republic	19.3N	68.9W	Earthqk	8.1	5.0	2.2	1 790
1946, Dec.	Japan, Honshu	33.0N	135.6E	Earthqk	8.1	6.6	2.7	1 362
1960, May	Central Chile	39.5S	74.5W	Earthqk	9.5	25.0	4.6	1 260
1969, Feb.	Indonesia, Makassar Straight	3.1S	118.9E	Earthqk	6.9	4.0	2	600
1976, Aug.	Philippines, Moro Gulf	6.26N	124.0E	Earthqk	8.1	4.5	2.3	2 349
1979, Dec.	Colombia, offshore	1.60N	79.4W	Earthqk	7.7	6.0	2.3	600
1992, Dec.	Indonesia, Flores Sea	8.48S	121.9E	Earthqk	7.8	26.2	4.7	2 500
1993, July	Sea of Japan	42.85N	139.20E	Earthqk	7.7	31.7	5	330
1994, June	Indonesia, Java	10.48S	112.84E	Earthqk	7.8	13.9	3.7	250
1998, July	Papua New Guinea	2.96S	141.93E	Earthqk	7.0	15.0	na	2 183
2004, Dec.	Indonesia, off W.Coast Sumatra	3.30N	95.98E	Earthqk	9.0	50.0	na	230 000
2006, July	Indonesia, Java	9.25S	107.41E	Earthqk	7.7	10.0	na	664

Notes and source: This table has been compiled from data in the NGDC data-base, see Table 11.5. 'Significant tsunamis' in this case are taken to be those causing a relatively large number of deaths. a Most tsunamis are caused by earthquakes – a few by volcanoes and landslides. b The water height above a reference level. Different reference levels may be used – see source for a short discussion. c Iida scale – see note in Table 11.5. na = not available.

11

Table 11.7 Tsunami events and fatalities, 1950–2007

Years	Events	Deaths
1950–1954	50	41
1955–1959	45	10
1960–1964	64	1 576
1965–1969	75	887
1970–1974	56	5
1975–1979	49	3 786
1980–1984	39	100
1985–1989	38	4
1990–1994	50	3 339
1995–1999	53	2 481
2000–2004	35	250 000
2005–2007	26	733
Totals	580	262 962
Av. per year	10	4 534

Note: Most tsunamis are caused by earthquakes, a few by landslides or volcanic activity. A tsunami of high magnitude need not necessarily involve a large number of deaths – many factors come into play.

Source: The table was compiled from data in the NGDC tsunami database. See Table 11.5.

Table 11.8 Landslide velocity classes

Class	Description	Velocity (mm s^{-1})	Nature of impact
7	Extremely rapid	>5000	Catastrophe of major violence; exposed buildings totally destroyed and population killed by impact of displaced material or by disaggregation of the displaced mass.
6	Very rapid	50–5000	Some lives lost because the landslide velocity is too great to permit all persons to escape; major destruction.
5	Rapid	0.5–50	Escape and evacuation possible; structure, possessions and equipment destroyed by the displaced mass.
4	Moderate	0.005–0.5	Insensitive structures can be maintained if located a short distance in front of the toe of the displaced mass; structures located on the displaced mass are extensively damaged.
3	Slow	5×10^{-5}–0.005	Roads and insensitive structures can be maintained with frequent and heavy maintenance work, if the movement does not last too long and if differential movements at the margins of the landslide are distributed across a wide zone.
2	Very slow	5×10^{-7}–5×10^{-5}	Some permanent structures undamaged, or if they are cracked by the movement, they can be repaired.
1	Extremely slow	<5×10^{-7}	No damage to structures built with precautions.

Source: Adapted from: Lee, E. M. and Jones, D. K. C. (2004). *Landslide Risk Assessment*. London: Thomas Telford. The table is based to a large part on the work of Cruden and Varnes.

Table 11.9 Significant landslides, 1919–2000

Type	Trigger	Date	Location	Deaths
Lahar	Volcanic	1919	Java, Indonesia	5 110
Landslide	Earthquake	Dec. 1920	Kansu, China	>200 000
Landslide	Earthquake	1920	Haiyuan, China	100 000
Landslide	Earthquake	Aug. 1933	Sichuan, China	6 800
Debris flow	n.i.	Dec. 1941	Huaraz, Peru	4000–6000
Landslide	Earthquake	1949	Khait, Tajikistan	12 000
Landslide	Earthquake	Aug. 1950	Assam, India	~30 000
Landslide	Rainfall	1958	Shizuoka, Japan	1 094
Landslide	n.i.	Oct. 1959	Minatilan, Mexico	5 000
Debris flow	n.i.	Jan. 1962	Nevados, Peru	4 000
Landslide	n.i.	Oct. 1963	Piave Valley, Italy	1 189
Landslide	Rainfall	1966	Rio de Janeiro, Brazil	1 000
Landslide	Rainfall	1967	Sierra des Araras, Brazil	1 700
Lahar	Earthquake	May 1970	Ancash, Peru	66 794
Landslide	n.i.	Sept. 1973	Choloma, Honduras	2 800
Lahar	Volcanic	Nov. 1985	Nevado del Ruiz, Colombia	>25 000
Landslide	n.i.	Apr. 1987	Cochancay, Ecuador	1 000
Landslide	Earthquake	Jan. 1989	Tajikistan	~10 000
Landslide	n.i.	June 1993	Nepal	3 000
Debris flow	Rainfall	Dec. 1999	Venezuala	30 000

Note: n.i. = no information.
Source: based on data in Galde, T., Anderson, M. and Crozier, M. J. (2005). *Landslide Hazard and Risk*. Chichester: John Wiley.

11

Table 11.10 Tornado intensity: Fujita Scale and Enhanced Fujita (EF) Scale

	Fujita Scale (1971)			EF-Scale 2007	
Level	Description	Wind Speed (km hr^{-1})	Level	Description	Wind Speed (km hr^{-1})
F0	Gale	64–116	EF0	Light damage	105–137
F1	Moderate	117–180	EF1	Moderate damage	138–178
F2	Significant	181–253	EF2	Considerable damage	179–218
F3	Severe	254–331	EF3	Severe damage	219–266
F4	Devastating	332–418	EF4	Devastating	267–322
F5	Incredible	419–512	EF5	Total destruction	>322
F6	Inconceivable	513–610			

Notes: The Fujita Scale is sometimes called the Fujita–Pearson scale in recognition of the contribution by Allen Pearson to the early development of the scale. The Enhanced Fujita Scale (EF-Scale) was introduced to deal with problems over the consistent use of the Fujita Scale, especially insofar as wind speed – usually at 3 second gusts – could not be applied meaningfully. The EF-Scale is applied retrospectively by assessment of the degree of damage caused by the tornado in relation to a set of damage indicators involving appropriate buildings, trees etc. It is, therefore, a proxy for wind speed. Details of the 28 damage indicators can be found on www.spc.noaa.gov/efscale and associated links (September 2008). Both scales are given here because they are cited in the literature.

Sources: The above website and Tobin, G. A. and Montz, B. E. (1997). *Natural Hazards, Explanation and Integration.* New York: The Guildford Press.

Table 11.11 Hurricane damage potential: Saffir–Simpson Scale

Level	Central pressure (10^{-3} bar)	Max. sustained wind speed (km hr^{-1})
1	>980	119–153
2	965–979	154–175
3	945–964	177–209
4	920–944	210–249
5	<920	>249

Source: As for Table 11.10.

Table 11.12 Annual frequency of hurricanes

Tropical basin	Hurricane[a]		Major hurricane[a]	
	Frequency	% of total	Frequency	% of total
Atlantic	5.7	12.1	2.2	10.9
NE Pacific	9.8	20.7	4.6	22.9
NW Pacific	16.8	35.5	8.3	41.3
N Indian	2.2	4.6	0.3	1.5
SW Indian (30–100°E)	4.9	10.4	1.8	9.0
Australian/SE Indian (100–142°E)	3.3	7	1.2	6.0
Australian/SW Pacific (142°E)	4.6	9.7	1.7	8.5
Global	47.3		20.1	

Notes: a Hurricane is a term often used for a strong tropical cyclone. A hurricane has a sustained wind speed of >33 m s^{-1} (119 km hr^{-1}), a major hurricane of >50 m s^{-1} (180 km hr^{-1}). The data above are averages mostly for the years 1970–2000.

Source: Based on data in Marks, F. D. (2003). Hurricanes. In *Encyclopedia of Atmospheric Sciences*, ed. J. R. Holton *et al*. Amsterdam: Elsevier/Academic Press.

Table 11.13 Meteorite hazard: estimated frequencies of impacts

Event type	Diameter of impactor	Energy[a] (megaton TNT)	Estimated average interval (a)	Impact probability per year	Reference source
Impact	10 m	0.1	20	5×10^{-2}	A
	50 m	10	10^3	10^{-3}	A
	100 m	100	5×10^3	2×10^{-4}	A
	200 m	600	4×10^4	2.5×10^{-5}	A
	277 m	1 000	6.3×10^4	1.6×10^{-5}	B
	597 m	10 000	2.41×10^5	4.1×10^{-6}	B
	1000 m	50 000	5×10^5	2×10^{-6}	C
	1287 m	100 000	9.25×10^5	1.1×10^{-6}	B
	5000 m	$\sim 7 \times 10^6$	2.5×10^7	4×10^{-8}	C
Tunguska[b]		$\sim 10\text{--}20$	$\sim 2000\text{--}3000$	$\sim 3\text{--}5 \times 10^{-4}$	C
K/T scale[b]	>10 km	$\sim 10^8$	$\sim 10^9$	$\sim 10^{-9}$	C

Notes: The estimated frequencies are derived from models based on the albedo distribution of near-Earth objects (NEOs). Errors are roughy $\pm 10\%$. a Impact energies are often given in units of megaton TNT (1 megaton TNT $= 4.18 \times 10^{15}$ J). The impact energy is primarily based on the mass of the impactor so the correlation with size is not exact. b The Tunguska event occurred in a remote part of Siberia in 1908. A K/T scale event is one similar in scale to the impact that occurred at the end of the Cretaceous at Chicxulub, Mexico and which is considered to have been a major cause of the mass extinction at that time. See Table 7.5 for major impact structures.

Sources: A Brown, P. et al. (2002). *Nature*, **420**, 294–296, B Morbidelli, A. et al. (2002). *Icarus*, **158**, 329–342. C Stuart, J. S. et al. (2004). *Icarus*, **170**, 295–311.

Appendix: SI Units, conversion factors and physical constants

This Appendix gives information on the standard units within the International System of Units (SI), together with selected factors for conversion from some non-SI units (some of which are still commonly used in the Earth Sciences) to SI units. It also gives selected physical constants. Numerous publications deal with SI units – a recommended one is that produced by the National Institute of Standards and Technology (U.S. Department of Commerce): *Guide for the Use of the International System Of Units (SI)*. NIST Special Publication 811. 2008 Edition. Tables A1–A3 are mainly based on this publication. It can be accessed via the web: http://physics.nist.gov/cuu/pdf/sp811.pdf

Table A.1 SI base and derived units

Physical quantity	Name	Symbol		
Base units				
Length	metre	m		
Mass	kilogram	kg		
Time	second	s		
Electric current	ampere	A		
Thermodynamic temperature	kelvin	K		
Amount of substance	mole	mol		
Luminous intensity	candela	cd		

Derived units	Name	Symbol	Definition	Equivalent form
Plane angle	radian	rad	1	m/m
Solid angle	steradian	sr	1	m^2/m^2
Frequency	hertz	Hz	s^{-1}	
Force	newton	N	$m\ kg\ s^{-2}$	J/m
Energy	joule	J	$m^2\ kg\ s^{-2}$	N m
Power, radiant flux	watt	W	$m^2\ kg\ s^{-3}$	J/s
Electric charge	coulomb	C	s A	
Electric potential difference	volt	V	$m^2\ kg\ s^{-3}\ A^{-1}$	W/A
Electric capacitance	farad	F	$m^{-2}\ kg^{-1}\ s^4\ A^2$	C/V
Electric resistance	ohm	Ω	$m^2\ kg\ s^{-3}\ A^{-2}$	V/A
Electric conductance	siemens	S	$m^{-2}\ kg^{-1}\ s^3\ A^2$	A/V
Magnetic flux	weber	Wb	$m^2\ kg\ s^{-2}\ A^{-1}$	V s
Magnetic flux density	tesla	T	$kg\ s^{-2}\ A^{-1}$	Wb/m2
Inductance	henry	H	$m^2\ kg\ s^{-2}\ A^{-2}$	Wb/A
Luminous flux	lumen	lm	cd sr	
Illuminance	lux	lx	$m^{-2}\ cd$	lm/m^2
Activity of radionuclide	becquerel	Bq	s^{-1}	
Amount-of-substance concentration		M	$10^3\ mol\ m^{-3}$	

Table A.2 SI prefixes

Multiple	Prefix	Symbol	Multiple	Prefix	Symbol
10^{24}	yotta	Y	10^{-1}	deci	d
10^{21}	zetta	Z	10^{-2}	centi	c
10^{18}	exa	E	10^{-3}	milli	m
10^{15}	peta	P	10^{-6}	micro	μ
10^{12}	tera	T	10^{-9}	nano	n
10^{9}	giga	G	10^{-12}	pico	p
10^{6}	mega	M	10^{-15}	femto	f
10^{3}	kilo	K	10^{-18}	atto	a
10^{2}	hecto	h	10^{-21}	zepto	z
10^{1}	deka	da	10^{-24}	yocto	y

Table A.3 Selected conversion factors – non-SI units to SI units

Convert from	To	Multiply by[a]
Area		
square foot (ft^2)	square metre (m^2)	9.290 304 E−02
square inch (in^2)	square metre (m^2)	6.4516 E−04
square mile (mi^2)	square metre (m^2)	2.589 988 E+06
square mile (mi^2)	square kilometre (km^2)	2.589 988 E+00
square yard (yd^2)	square metre (m^2)	8.361 274 E−01
Electricity and magnetism		
gauss (Gs, G)	tesla (T)	1.0 E−04
Energy		
calorie$_{th}$ (cal$_{th}$)	joule (J)	4.184 E+00
electron volt (eV)	joule (J)	1.602 176 E−19
erg (erg)	joule (J)	1.0 E−07
ton of TNT (energy equivalent)	joule (J)	4.184 E+09
Force		
dyne (dyn)	newton (N)	1.0 E−05
kilogram-force (kgf)	newton (N)	9.806 65 E+00
poundal	newton (N)	1.382 550 E−01
pound-force (lbf)	newton (N)	4.448 222 E+00
Length		
ångstrom (Å)	metre (m)	1.0 E−10
astronomical unit (ua)	metre (m)	1.495 979 E+11
foot (ft)	metre (m)	3.048 E−1
inch (in)	metre (m)	2.54 E−02
light year (Ly.)	metre (m)	9.460 73 E+15
mile (mi)	kilometre (km)	1.609 344 E+00
yard (yd)	metre (m)	9.144 E−01
Mass and mass divided by area		
carat, metric	gram (g)	2.0 E−01
ounce (avoirdupois) (oz)	gram (g)	2.834 952 E+01
pound (avoirdupois) (lb)	kilogram (kg)	4.535 924 E−01

Table A.3 (*cont.*)

Convert from	To	Multiply by[a]
Mass and mass divided by area (*cont.*)		
ton, long (2240 lb)	kilogram (kg)	1.016 047 E+03
tonne (t)	kilogram (kg)	1.0 E+03
pound per square foot (lb/ft^2)	kilogram per square metre ($kg\,m^{-2}$)	4.882 428 E+00
Pressure		
atmosphere, standard (atm)	pascal (Pa)	1.013 25 E+05
bar (bar)	pascal (Pa)	1.0 E+05
kilogram-force per square metre (kgf/m^2)	pascal (Pa)	9.806 65 E+00
torr (Torr)	pascal (Pa)	1.333 224 E+02
Temperature		
degree Celsius (°C)	kelvin (K)	$t\,°C + 273.15$
degree Fahrenheit (°F)	kelvin (K)	$(t\,°F + 459.67)/1.8$
Viscosity		
poise (P)	pascal second (Pa s)	1.0 E−01
stokes (St)	metre squared per second ($m^2\,s^{-1}$)	1.0 E−04
Volume		
cubic inch (in^3)	cubic metre (m^3)	1.638 706 E−05
cubic yard (yd^3)	cubic metre (m^3)	7.645 549 E−01
cubic mile (mi^3)	cubic metre (m^3)	4.168 182 E+09
gallon (Imperial) (gal)	litre	4.546 09 E+00
gallon (US) (gal)	litre	3.785 412 E+00

Note: *a* The letter E in this column stands for 'exponent', with the following digits indicating the power of 10 by which the preceeding number is multiplied.

Table A.4 Selected physical constants

Quantity	Symbol	Value
Atomic mass constant	m_u	$1.660\ 538\ 782 \times 10^{-27}$ kg
Avogadro constant	N_A or L	$6.022\ 141\ 79 \times 10^{23}$ mol^{-1}
Bohr magneton	μ_B	$927.400\ 915 \times 10^{-26}$ J T^{-1}
Boltzmann constant	k	$1.380\ 6504 \times 10^{-23}$ J K^{-1}
Electron magnetic moment	μ_e	$-928.476\ 377 \times 10^{-26}$ J T^{-1}
Electron mass	m_e	$9.109\ 382\ 15 \times 10^{-31}$ kg
Electron volt	eV	$1.602\ 176\ 487 \times 10^{-19}$ J
Faraday constant	F	$96\ 485.3399$ C mol^{-1}
Gas constant	R	$8.314\ 4725$ J K^{-1} mol^{-1}
Neutron mass	m_n	$1.674\ 927\ 211 \times 10^{-27}$ kg
Newtonian constant of gravitation	G	$6.674\ 28 \times 10^{-11}$ m^3 kg^{-1} s^{-2}
Planck constant	h	$6.626\ 068\ 96 \times 10^{-34}$ J s
Proton mass	m_p	$1.672\ 621\ 637 \times 10^{-27}$ kg
Speed of light (vacuum)	c	$299\ 792\ 458$ m s^{-1}

Source: National Institute of Standards and Technology (U.S. Department of Commerce). *Fundamental Physical Constants.* Web address: http://physics.nist.gov/cuu/index.html Last updated December 2003, accessed August 2008.

Index